Fundamentals of
Soil Science
Second Edition

Shivanand Tolanur MSc (Agri), PhD

Professor and Head
Department of Soil Science and
Agricultural Chemistry
College of Agriculture
University of Agricultural Sciences
Vijayapur, Karnataka (India)

CBSPD

CBS Publishers & Distributors Pvt Ltd

New Delhi • Bengaluru • Chennai • Kochi • Kolkata • Lucknow • Mumbai
Hyderabad • Jharkhand • Nagpur • Patna • Pune • Uttarakhand

Fundamentals of Soil Science

Second Edition

ISBN: 978-93-87085-08-4

©Publisher

Second Edition: 2018
Reprint: 2021, **2024**

First Edition: 2004

Published by **Satish Kumar Jain** and produced by **Varun Jain** for

CBS Publishers & Distributors Pvt Ltd
4819/XI Prahlad Street, 24 Ansari Road, Daryaganj, New Delhi 110 002, India.
Ph: 011-23289259, 23266861   Website: www.cbspd.com
                                                       e-mail: delhi@cbspd.com

Corporate Office: 204 FIE, Industrial Area, Patparganj, Delhi 110 092
Ph: 011-4934 4934   Fax: 011-4934 4935
                                          e-mail: publishing@cbspd.com; publicity@cbspd.com

Branches
_____

- **Bengaluru:** Seema House 2975, 17th Cross, KR Road, Banasankari 2nd Stage, Bengaluru 560 070, Karnataka, India
  Ph: +91-80-26771678/79   Fax: +91-80-26771680   e-mail: bangalore@cbspd.com
- **Chennai:** 7, Subbaraya Street, Shenoy Nagar, Chennai 600 030, Tamil Nadu, India
  Ph: +91-44-26680620, 26681266   Fax: +91-44-42032115   e-mail: chennai@cbspd.com
- **Kochi:** 42/1325, 1326, Power House Road, Opp KSEB, Power House, Ernakulum Kochi 682 018, Kerala, India
  Ph: +91-484-4059061-65,67   Fax: +91-484-4059065   e-mail: kochi@cbspd.com
- **Kolkata:** 147, Hind Ceramics Compound, 1st Floor, Nilgunj Road, Belghoria, Kolkata-700056, West Bengal, India
  Ph: +033-25633055, 033-25633056   e-mail: kolkata@cbspd.com
- **Lucknow:** Basement, Khushnuma Complex, 7 Meerabai Marg (Behind Jawahar Bhawan), Lucknow-226001, UP, India
  Ph: +0522-4000032   e-mail: tiwari.lucknow@cbspd.com
- **Mumbai:** PWD Shed, Gala no 25/26, Ramchandra Bhatt Marg, Next to JJ Hospital Gate no. 2, Opp. Union Bank of India,
  Noorbaug, Mumbai-400009, Maharashtra, India
  Ph: 022-66661880/89   e-mail: mumbai@cbspd.com

Representatives
_____

- Hyderabad   0-9885175004   • Jharkhand   0-9811541605   • Nagpur   0-8692091830
- Patna   0-9334159340   • Pune   0-9664372571   • Uttarakhand   0-9716462459

Printed at SRK Graphics, Delhi (India)

Affectionately Dedicated to
My beloved Father
Late Sri IRAPPA
Wife PREMA and
Daughters SHRUTI, SUSHMA

CONTENTS

CONTENTS

DEFINITION OF SOIL

A. Definition : A definition is a descriptive statement conveying the properties and characteristics of an object in nature, or the meaning of an idea, phenomenon, or a term.

As much as possible, a definition should be scientific, i.e., true to established facts. And lastly, a definition must be brief and expressed in clear and simple language.

Definition of Soil

In early days, soil was defined as the upper loose layer of the earth suitable for plant growth

1) As late as 1917, Ramann defined soil as "the uppermost layer of the solid crust of the earth; (Crust = A surface layer on soils, ranging in thickness from a few millimeters to perhaps as much as 3 cm, that is much more compact, hard and brittle when dry than the material immediately beneath it); it consists of rocks that have been reduced to small fragments and have been more or less changed chemically, together with the remains of plants and animals that live on it and use it". Such definition does not distinguish between soil and loose rock material.

2) Hilgard defined soil as "the more or less loose and friable material in which, by means of their roots, plants may or do find a foot hold and nourishment, as well as other conditions of growth" (This is a purely agronomic or plant physiological definition).

3) Wahnschaffe, Mitscherlich and other western European soil wokers which reads as follows :- "Soil is a mixture of pulverized solid particles, water and air which may serve as

1

a carrier of available plant food materials for growth.

These definitions may be summed up as the concept that the soil is just a medium for plant growth. Under these definitions, the sand and solution cultures are also soils.

4) Dokuchaev (1900) defined soil as "the surface and adjoining horizons of parent material (irrespective of the kind) which have undergone, more or less, a natural change under the influence of water, air and various species of organisms-living or dead; this change is reflected, to a certain degree in the composition, structure, and colour of the products of weathering".

Jenny (1941) soil is naturally occurring body that has been evolved owing to combined influence of climate and organisms, acting on parent material, as conditioned by relief over a period of time.

5) It was Marbut who made a definite step forward in defining soil in terms of soil characteristics instead of soil forming processes: "The soil consists of the outer layer of the earth's crust usually unconsolidated ranging in thickness from a mere film to a maximum of somewhat more than ten feet which differs from the material beneath it, also usually unconsolidated, in colour, structure, texture, physical constitution, chemical composition, biological characteristics, probably chemical processes, in reaction and morphology".

Any modern definition of soil – would place pedology on the same level with the other natural sciences – the statement that the soil is "an independent natural body" Pedology – As an independent science originated in Russia – One of the subject pedology which deals about soil. Pedology pedo-ground logus-science. Nothing but soil science. Soil is an independent natural body. A point might also be raised about the phrase, "the outer layer of the earth's crust", which conveys the earlier geological concept instead of outer layer it should be surface layer. After critical analysing the definition of soil physical, morphological properties. Then soil may thus be defined as follows : "The soil is a natural body of mineral and organic constituents, differentiated into horizons, of variable depth, which differs from the material below in morphology, physical makeup, chemical properties and composition, and biological characteristics".

Morphology is not a science : It is an aid to science, one of the methods used in scientific investigations. It is an art which requires keen observation and ability to describe and record in words and drawings an object studied. The primary aim of morphology is description.

Soil morphology – description of the soil body, its appearance, features, and general characteristics as expressed in the profile of a virgin soil e.g. Podzol [Profile layer/morphological term (kind of parent material)].

Soil Anatomy : Soil profile consists of 3 or 4 genetic layers (i) humus decay – accumulative (ii) eluvial (iii) illuvial (iv) parent material immediately below the illuvial horizon. The successive steps in examining and describing the morphology of the soil are applicable to all profiles irrespective of the origin of the parent material or state of the soil body.

The State of Soil body –

The following characteristics are to be looked for : (1) colour (2) constitution (3) habitus of the profile (4) depth of profile and thickness of respective horizons (5) texture of soil material (6) structure (7) concretions and foreign instrusions and (8) miscellaneous observations. These characteristics are scrutinized first of all by 4 of the natural senses.

(1) sight (2) touch (3) smell and (4) sometimes taste

SOIL SCIENCE AND ITS BRANCHES

In introducing the subject of Pedology. It is the synonym of soil science. Soil science is an independent science. Pedology originated in Russia. Russian scientist Dokuchaev he worked and contributed classical researches to soil science (Pedology). Russian language is a problem outside Russia.

In USA scientist Marbut he gave much interest in modern scientific idea about soils.

Dr. Glinka, student of Dokuchaev contributed more ideas towards the soil science. In Russia, Soil Science is treated as one of the natural science such as Zoology & Botany.

Two disciplines like Chemistry and Physics are involved in the soil studies like Soil Chemist, and Soil Physicist. Both Soil Physicist and Soil Chemist devoted their attention to the soil material rather than to the soil body. (Chemistry – soil, acidity, liming, soil solution, soil colloids and its properties related to chemical) (Physics – sp. gravity, volume weight, capillary pore space, moisture regime, texture, structure of soils and mechanical analysis) may be easily applied.

An analysis of the evolution of the soil science as a distinct scientific discipline would be incomplete without mention of geology, petrography and mineralogy makes a strong appeal to the students of soils.

Soil form the uppermost layer of the earth's mantle, and very frequently, are formed in situ, retaining some of the rock fragments and minerals of the underlying geological formation. Although geology served as a great aid in the recognition and evaluation of certain soil properties, it was inadequate to explain the complex processes of soil formation.

It remained for pedology to solve the difficulty. By recognizing the soil as an independent natural body in relation to the processes responsible for its formation, pedology had justified its claim to an independent status of a scientific discipline.

Relationship between Microbiology and Soils

An elemental constituent of the soil is its organic matter, the product of biological agencies in the process of soil formation. Decomposition and synthesis reactions brought about by microbes in the soil are important in elucidating some fundamental properties of the soil.

Microbiologists gave an idea about the processes of organic matter transformation in the soil, ammonification, nitrification and nitrogen fixation. But microbiologists have not investigated the relation of microbial activities to the processes of soil formation. Their investigation dealt primarily with soil mass as medium for microbes and the effect of their activities on crop production. In this respect, soil microbiology gave some thing to agronomy and plant physiology rather than soil science. In general it is less to pedology.

Agronomy and Soils

The agronomic point of view stresses the fertility of the soil expressed in yields per acre. Tillage and other soil management practices, compensation for crops taken off, and systems of cropping are the criteria which guide by the agronomist. For agronomist soil is only supporting material to the plants. Only he is giving an attention towards cultivated soil, even though he realizes the potentialities of the virgin soil. Agronomists are giving much importance towards yields of crops. But they are not giving much consideration towards physical and chemical biological properties of soils.

Pedology and the pure and Applied Sciences

Achievements of chemists, geologists, agronomists they have studied the soil their own purpose.

Pedology – (Soil Science) presents the soil as a unit in nature; it deals with its origin, formation and distribution through a study of its constitution and its life and dynamics.

Soil Science (Pedology) occupies an intermediate position between the science of animate and inanimate nature.

Soil Science mobilizes the findings of the natural science and applies them in elucidating the concept of the soil in relation to the biosphere, hydrosphere and lithosphere.

Relation of Pedology (Soil Science) to other Scientific disciplines.

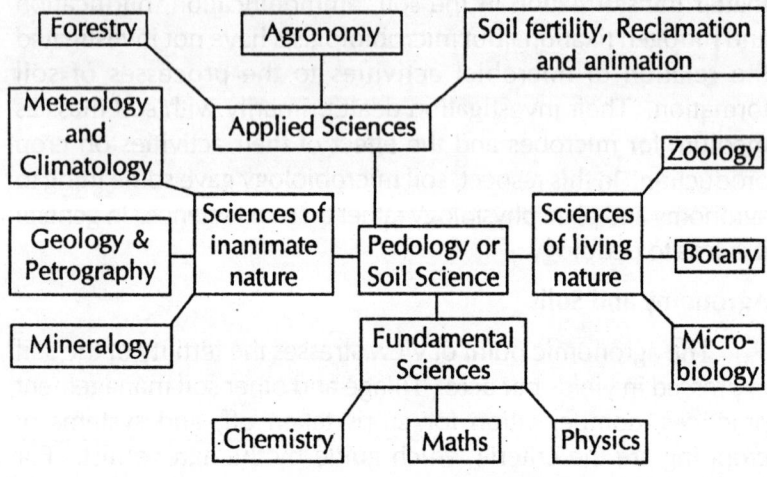

Soil – Primary Natural Object, provides energy, home and health food and fibre for the every survival of mankind. "Pedology" as a synonym for soil science (Fallow, 1862, Glinka 1927, Sigmond 1938, Robinson, 1949) What is Pedology? (Greek: Pedon Ground, or earth : logos = Science)

Pedology or soil science is the applied scientific discipline at the meeting point of physical (e.g. Physics; Chemistry: Maths); geological (e.g. mineralogy, geology, geography) : Biological (e.g. botany, zoology, microbiology) and agricultural (e.g. soil technology and crop production) sciences (Fig. 1).

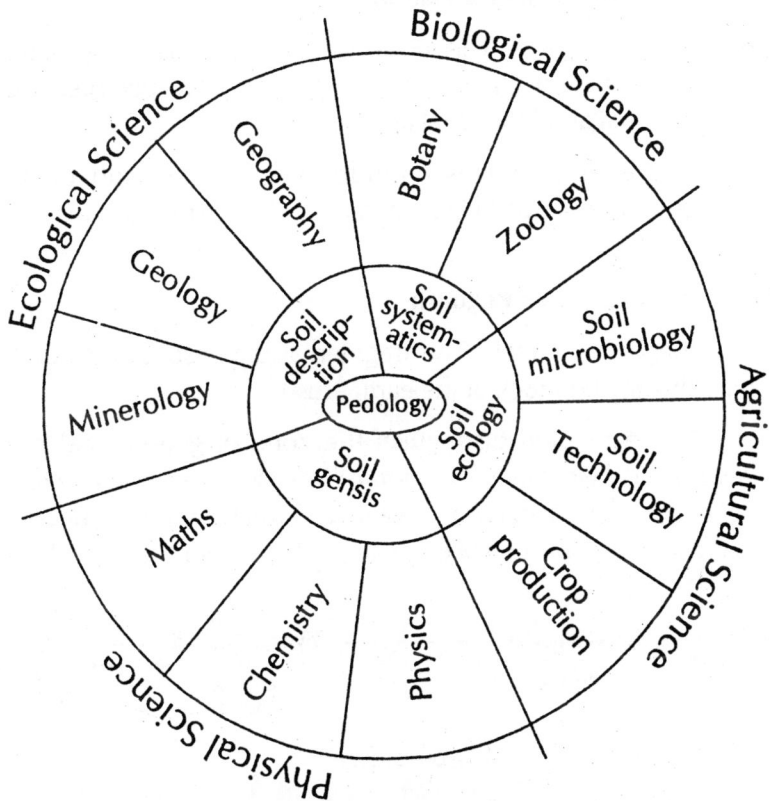

Fig. 1. Important branches of Pedology

The important branches of Pedology are as follows :

(a)   Soil genesis : This is the study of the formation and development of soil through physical, geochemical and geobiochemical weathering and transformation of the regolith by the combined activities of the environmental pedagenic factors over time and space.

(b) Soil description : This comprises the inventory of soil description and characterization through soil survey, mapping and interpretation.

(c) Soil Systematics : This includes soil correlation, classification into different classes or texa according to pedogenetic, regional and functional aspects.

(d) Soil ecology : This includes the study of soil in the environment of living, macro, and micro organisms and mankind.

Concept of Soil and Regolith

(Regolith : Greek. The fragmental unconsolidated debris mantling the bed rock of the earth crust)

In the regolith distinguish the zone of geophysical and geochemical weathering as "not soil" that is parent material from zone of geo-biochemical or pedological weathering as living soil" in the Katamorphic belt of earth's crust (Polynovo 1937, VanHise 1904)

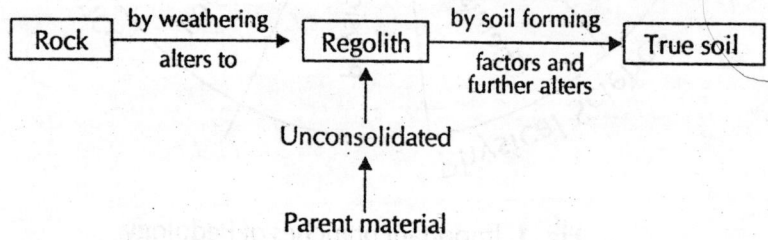

Soil Volume

Volume composition of loam surface soil when conditions are good for plant growth. We stated that where the regolith meets the atmosphere, the worlds of air, rock, water and living things are intermingled (Fig. 2).

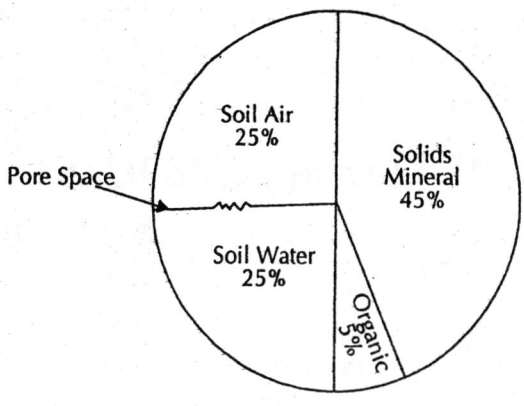

Fig. 2. Volume composition of soil

In fact, the four major components of soil are air, water, mineral matter and organic matter. The relative proportions of these 4 components greatly influence the behaviour and productivity of soils. In a soil, the 4 components are mixed in complex pattern.

Composition of soil : From the physical stand point; soil is composed of solid phase, liquid phase and gaseous phase in varying proportions (by volume).

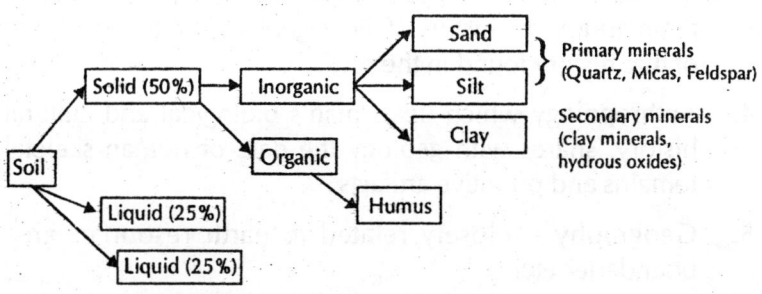

Approximate composition of surface soil.

9

SOIL FORMING ROCKS AND MINERALS

Geology is the science of the earth, its composition and structure, its history and its past plant and animal life.

Geology it is independent science, and it also rests upon a foundation of Astronomy, Chemistry, Physics and Biology.

Geology – closely related to anthropology, geography and economics

1. Astronomy : We have to study the earth – its origin, its place in Solar System and its relation to the universe as a whole. These are matters of astronomy.

2. Physics and Chemistry – Study of the composition and structure of the earth involves questions of chemistry and physics.

3. Geology & Biology – The evolution of living things is a common concern of geology and biology, the justification for including the history of life in geology is that the records of that history found in the rocks.

4. Anthropology which treats man's biological and cultural history, shares with geology the date of human skeletal remains and primitive artifacts.

5. Geography – closely related to earth resources area boundaries etc.

6. Economics – Available source like iron, oil, coal, gold and other earth resource.

7. Mathematics is the fundamental to every physical science.

Earth

The sun is the center of the solar system, in which solid bodies that revolve about the sun are called planets. Nine planets in the order of their proximity to the sun are Mercury, Venus, Earth, Mars, Jupiter, Saturn, Uranus, Neptune and Pluto.

Sun		Distance from sun million miles
1.	Mercury	36
2.	Venus	67
3.	Earth	93
4.	Mars	142
5.	Jupiter	483
6.	Saturn	886
7.	Uranus	1780
8.	Neptune	2790
9.	Pluto	3670
	Moon	98

Shape of Earth : The earth is an oblate spheroid nearly spherical but slightly flattened at the poles and bulging at the equator. The surface of earth is smooth.

Size of the earth : The earth has a polar diameter of about 7900 miles and an equatorial diameter of about 7927 miles. Its circumference around the equator, is approximately 24,900 miles. The area of earth's surface is about 197 million square miles of which about 71% is covered by ocean. The volume of the earth is a little more than 250 million cubic miles, and its mass has been estimated at about 6,600 quintillion tons (6,600,000,000,000,000,000,000, tons)

Major divisions of Earth (Fig. 3)

There are 3 spheres corresponding to the 3 states of matter (Solid, liquid & gas) which constitute the earth.

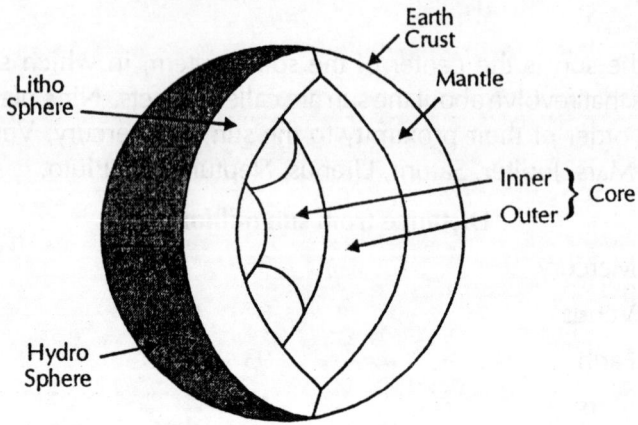

Fig. 3. Major divisions of earth

Solid (1) The solid zone is the Lithosphere Liquid (2) The incomplete covering of water forming seas and oceans is the Hydrosphere Gas (3) Gases envelop over the earth's surface is the Atmosphere

1. Atmosphere : The envelop of air that covers both the lithosphere and hydrosphere is called atmosphere. Component or composition of earth

    Nitrogen : 78.084% (Main component) Oxygen : 20.946% the gas significant to man) Argon : 0.934% Carbondioxide : 0.033%

    In addition, inert gases, such as neon, helium, krypton and xenon, are present. Water vapours present in the air vary in amounts at different places and times.

2. Hydrosphere : It is a sphere of water surrounding the earth. Water covers three fourth of the earth's surface. Most of it lies within the ocean basins, it also appears on the surface of land in the form of river, ponds, lakes and as ground water.

3. Lithosphere : It is solid surface (continents, ocean, basin, plains, plateaus and mountains, valleys, sand dunes, lava flows and fault scraps) and interior of the earth consists of

rocks and minerals. It is covered by gaseous and watery envelops. It amounts to 93.06% of the earth's mass.

Interior of the Earth

The earth ball consists of 3 concentric rings : Crust, Mantle and Core (inner and outer) (Fig. 3)

Crust : 5-56 km thick consists of rocks

Mantle : 2900 km in thickness, contains mixed metals and silicates and ultra basic rocks

Core : The innermost portion of the earth it is 3,500 km in thickness, comprises metals, such as nickel and iron

Composition of the Earth's crust

Most of the hard, naturally formed substances of the earth is referred to as Rock. Rock is composed of elements, which in turn are made up of atoms. Out of 106 elements known, 8 are sufficiently abundant as to constitute 98.6% (by weight) of the earth's crust (up to 16 km).

The two elements occuring in greatest abundance are non-metallic (Oxygen and silicon) and comprise nearly three fourth of the total composition of the crust. The other 6 elements are metals. The commonly known as valuable metals like gold, silver, nickel, copper and zinc are of rare occurance.

Average composition of the Earth's crust (% by wt)

Non metallic	O^{2-}	46.60% }	(74.32%) i.e. ¾ of the
	Si^{4+}	27.72% }	total
Metals	Al^{3+}	8.13% }	
	Fe^{2+}	5.00% }	
	Ca^{2+}	3.63% }	(25.68%) i.e. ¼ of the
	Na^+	2.83% }	total
	K^+	2.59% }	
	Mg^{2+}	2.09% }	
	Others	1.41% }	

The materials of the Earth's crust fall into two principal categories, namely

(1) Minerals and (2) Rocks

Rocks – The rocks are generally of 2 or more minerals. On the basis of their mode of origin, the rocks are divided into 3 main groups.

(1) Igneous (2) Sedimentary and (3) Metamorphic

They are the outcome of the geological forces or processes which operate on the surface and the interior of the earth to bring about changes to produce rocks.

Relative abundance of rocks in Earth's Crust

The composition of the upper 5 km of the earth's crust is as follows.

1. Sedimentary	Shales	52%	
	Sandstone	15%	74%
	Lime stones & dolamite	7%	
2. Igneous Rocks	Granite	15%	18%
	Basalt	3%	
3. Others		8%	8%
		100	100

It is evident from the data given above that 5 kinds of rocks occupy more than 90% of the total continental area. However, the composition of Earth's crust as a whole differs significantly from one described above

- Igneous Rocks   95%

                                           Shales   (4.0%)
- Sedimentary Rocks 5%   {   Sand stone   (0.75%)
                                           Lime stone   (0.025%)

Although sedimentary rocks form only 5% of the total Earth's crust, yet they are important as they occur to the extent of 74% (almost ¾) at or near the surface of the earth (upper 5 Km). The figures given above also show that as we go deeper, we find the predominance of igneous rocks.

Rock forming minerals - their formation, characteristics and classification

The rocks which form the earth are made of minerals

Definition : Minerals

Minerals are solid substances and composed of atoms having an orderly and regular arrangement. Precisely, a mineral is a naturally occurring, homogeneous element or inorganic compound that has a definite chemical composition and a characteristic geometric form.

Most of minerals consists of 2 or more elements combined to form a compound, such as gypsum – $CaSO_4$, $2H_2O$, Olivine $(Mg, Fe)_2 SiO_4$ or feldspar – $KAl Si_3O_8$

Some minerals, however, may consist of only one of the naturally occurring chemical elements, which may be a metal, such as copper, iron, calcium, or non metal such as carbon, sulphur, silicon. The formula SiO_4 means the silicon and oxygen ions are in the ratio of 1:4.

Being homogeneous, a mineral exhibits uniformity on its physical properties which are used for its identification.

Physical properties for Mineral Identification

The properties or characteristics by which a mineral can be identified are numerous. But some practical properties used for mineral identification are :

1.	Colour	2.	Streak	3.	Striation
4.	Hardness	5.	Lustre	6.	Transparency
7.	Specific gravity			8.	Tenacity
9.	Cleavage and Fracture				
10.	State of Aggregation and crystal form.				

1. Colour : Each mineral has a definite colour or gradation of colours which aid in its identification. In nature, the colour of minerals is variable; a mineral may have more than one colour depending upon its chemical composition. The colours of a few important rock forming minerals are :

Minerals	Colour
1. Quartz	Colourless
2. Feldspars, Calcite, Dolomite Gypsum, Kaolinite, Muscovite	White to pale
3. Iron pyrite (called "Fools Gold")	Yellow/Golden
4. Olivine, Serpentine, Horn blende	Greenish
5. Garnet	Reddish-brown
6. Biotite, Horn blende, Augite, Haematite Magnetite, Graphite	Black
7. Orthoclose	Pink or Flesh coloured

2. Streak : Although colour may vary, the streak i.e. fine powder of the mineral, representing its true colour, is of greater reliability. Streak is produced and determined by rubbing the specimen on a piece of unglazed porecelain plate, called a streak. ex : Haematite - black in colour, give red brown colour after streak.

3. Striation : The parallel thread-like lines or narrow bands running across the surfaces of a mineral are called striations. These are reflections of the internal arrangement of atoms of crystals. These are clearly observed on crystals of quartz, feldspars and pyrite.

4. Hardness : The resistance of a mineral to scratching is known as hardness. The hardness is expressed in Mho's scale and indicated by numbers (1 to 10). Minerals vary widely in their hardness; talc is the softest and diamond the hardest mineral known

Hardness (Mho's scale)	Mineral Substances	Test
1.	Talc	Scratches by finger nail
2.	Gypsum	Just scratches by a finger nail
3.	Calcite	Scratches not easily by a copper coin piece
4.	Flourite	Scratches easily by a steel knife
5.	Apatite	Just scratches by a knife
6.	Feldspar eartho clase	Scratches soft glass
7.	Quartz	Scratches glass easily
8.	Topaz	Scratches glass but not hard enough to be used as grinding material
9.	Carborundum or Corundum	Very hard and used as grinding material for all minerals
10.	Diamond	hardest mineral known

5. Lustre : It is the general appearance of a mineral in reflected light. It is the characteristic of each mineral

   For e.g. : Iron minerals have metallic lustre

   Clay minerals – have dull lustre

   Micas having a shining lustre

   Quartz – a vitreous lustre (glass) calcite

6. Transparency : It is the degree of penetration of light through a mineral. Some minerals may be Transparent : e.g. mica, Translucent : e.g. Quartz/Opaque, eg. Pyrite, Magnetite etc.

7. Specific gravity

   It is the ratio between the weight of a mineral or substances to the weight of an equal volume of water

   The specific gravity of a mineral increases with the mass number of its constituent elements and with closeness of packing of these elements in their crystalline form.

   The specific gravity of different minerals varies, for instance, quartz – 2.65, haematite – 5.2. Most of the rock forming minerals have a specific gravity around 2.7

   The minerals are divided into two groups based on their specific gravity.

1. Heavy minerals : have specific gravity greater than 2.85, e.g. Pyroxene, Amphiboles, Garnet, Zircon etc.

2. Light minerals have specific gravity below 2.85 e.g. Quartz, micas, feldspars etc.

8. Tenacity : The resistance that a mineral offers to breaking, crushing or binding is known as tenacity. Minerals may be brittle, malleable, flexible or elastic.

9. Cleavage and Fracture : The tendency of a mineral to split in certain preferred directions along smooth plane surfaces is called cleavage.

The cleavage planes are governed by the internal arrangement of atoms and the directions in which the atomic bonds are relatively weak, as in micas and feldspars.

Fracture : on the other hand, is the property of a mineral to break along an irregular surface, not connected with crystalline form, as in glass and quartz.

10. State of aggregation and crystal form : Almost all the rock forming minerals have got crystalline structure. The specific atomic arrangement of the minerals (as in crystal) is called its crystal form.

A crystal may be defined as a substance with regular geometric faces and definite structure. The angle between the faces is always constant for a particular crystal.

Occurrence of rock forming minerals

In more than 2000 known minerals only a few occur in abundance in the earth crust.

Formation of Minerals

When the molten magma solidifies, the different elements present herein freely arrange themselves in accordance with attractive forces and geometric form.

Table : Important rock-forming minerals (Primary and Secondary silicates) and their relative abundance

Minerals (arranged in order of their crystallization)	Important constituents other than Si and O	% distribution
A. Primary minerals Ferromagnesians		
1. Ortho or Ino-silicates		
-Olivines	Fe, Mg	
- Pyroxenes	Ca, Na, Fe, Mg	16.8%
-Amphiboles etc.	Ca, Na, Fe, Mg, Al, OH	

2. Phyllosilicates		
- Biotite	K, Fe, Mg, Al, OH	
- Muscovite	K, Al, OH	3.6%
Non-ferromagnesians		
3. Tecto silicates		
Feldspars such as		
Anorthite	Ca, Al	
Albite	Na, Al	61.0%
Orthoclase	K, A1	
Quartz		11.6%
B. Secondary clay minerals	Na, K, ca, Mg, Fe, Al, OH	6.0%
C. Others		1.0%
	Total	100.00

Geometrically, it is possible to arrange only 4 oxygen anions (with radius 1.32A°) around a central silicon cation (with radius of 0.42A°) so that all are touching each other. This is the arrangement of a tetrahedron (Fig. 4).

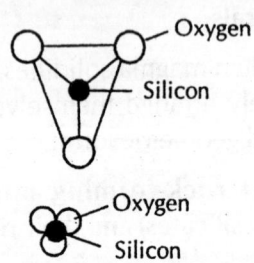

Fig. 4. Arrangement of tetrahedron

The silicate tetrahedron is the fundamental building block of all the silicate minerals of the earth's crust. The amount of charge carried by silicon ion is 4^+ and by oxygen is 2^-. In order to attain neutrality, one silicon (4^+) ion would combine with two oxygen ions $(2 \times 2^-)$ to form SiO_2, but geometrically stable structure is formed when 1 silicon combines with 4 oxygen ions to form tetrahedron $(SiO_4)^{4-}$ which carries a net negative charge of 4.

In nature, the geometry and valency constraints are reconciled; first, by linking together tetrahedron so that oxygen ions are shared between neighbouring silicon thus reducing negative charge deficit. Second, by making use of the positive charges of other metal cations to balance the negative charge. Both these occur together to produce a neutral mineral.

As a result of this, the basic tetrahedral arrange themselves in an orderly manner according to a fixed plan, forming different minerals which are called primary minerals. The minerals formed due to the weathering of pre-existing primary minerals are called secondary minerals. In the absence of a crystalline form, the material is termed as non-crystalline e.g. allophane chalcedony.

Silicates

A) Primary Minerals

The primary minerals are those which are formed due to the crystallization of the molten magma. Already seen that Earth's crust contains dominant amount of oxygen (46.60%) followed by silicon (27.72%). There would be a greater tendency for silicon and oxygen to combine to form the basic compound, called silicon-oxygen tetrahedron $(SiO_4)^{4-}$. This explains the dominance (> 90%) of silicate minerals in the earth's crust.

Tetrahedral links :

The silicates that are made up of individual silicon terahedra $(SiO_4)^{4-}$, alternating with +vely charged metal ions (Fe, Mg), are orthosilicates of which olivine is a typical example.

Phyllosilicate : A more complex linkage. All tetrahedral share 3 oxygen ions with neighbouring tetrahedral. The sheet structure are phyllosilicates mica best example.

Tectosilicate tetrahedral share with 4 oxygen ion. Such 3 dimensional structure are termed as tectosilicates e.g. Quartz & feldspar.

1.   Quartz : Very densely packed and occurs in a high degree

of purity. Quartz is strongly resistant to physical and chemical weathering (due to structure itself densely packed) electrically neutral and prevents any form of substitution. After decompose to form clay.

2. Feldspars : 61% of feldspar mineral that are found in the earth's crust. They are non-ferromagnesian minerals and act as store house of Na, Ca, K, & many trace element in soils.

B. Secondary Minerals

The secondary minerals are formed at the earth's surface by weathering of pre-existing primary minerals.

During weathering, water accompanied by CO_2 from the atmosphere plays an important role in processes, such as hydrolysis, hydration, and solution. As a result the primary minerals are altered or decomposed.

Feldspar + Water – Claymineral + Cations + anions + soluble silica

In the weathering many elements are released into solution; a part of which may be used as a source of plant nutrients, a part may be leached out into the ground water, CO_2, H_2O combine to form secondary minerals (like Kaolinite, montmorillaute illite etc.) secondary minerals are clay minerals and iron & aluminium oxides.

Other secondary minerals observed in soils especially in arid and semi-arid (dry) regions are gypsum, calcite, attapulgite and aguite.

Some important secondary minerals

Silicates – clay minerals : hydrous aluminosilicate with layer structure similar to micas eg. illite, montmorillonite, kaolinite etc.

Non-silicates

1. Oxies; hydroxides or oxyhydroxide of Si, Al & Fe

    (i) Haematite Fe_2O_3 (ii) Limonite, $FeO (OH)_n H_2O$ (iii) Gibbsite $Al(OH)_3$

2. Carbonates : (i) Calcite $CaCO_3$ (ii) Dolomite $CaMg (CO_3)_2$

3. Sulphates : (i) Gypsum $CaSO_4 2 H_2O$

4. Phosphates : (i) Apatite $Ca_3 (PO_4)_3$

Clay Minerals : Silicate secondary minerals found in the clay fraction ($> 2\mu$) of soils. The most commonly observed are layer silicates (illite, montomorillonite, chlorite, vermiculite, kaolinite). Besides O, OH, Al and Si they contains Mg, Fe and K in large amounts. They are variable in colour (white, grey, light yellow) depending on their chemical composition. In soils, the clays and oxihydrates of iron which form coatings on mineral grains impart shades of yellow, brown or red colour to soils.

The clay minerals carry a significant negative electrical charge on their surfaces. In some cases the groups of sheets are not firmly bounded together and water molecules can enter in their crystal lattice. This can cause considerable swelling due to change in soil moisture content. This is the case in Vertisols (Black cotton soils) of India where deep and wide cracks on the surfaces are suggestive of the shrink swell characteristics of soil clays.–ve electrical charge on the clay surfaces, the cations are attracted to regions of electrical charge around the clay minerals.

The amount of –ve charge varies depending upon the type of clay mineral and it is referred to as the CEC. This exchange there is always a balance between the concentrations of cations in soil water and those adsorbed on the surface of the particles.

Distribution of Minerals (Silicate Minerals)

While primary minerals are observed in all rocks and in sand and silt fractions of soils. The secondary minerals dominantly occur in clay fractions of almost all soils and in sedimentary rocks especially shales.

The kind and proportions of mineral(s) found in a soil depend on the kind of parent material and the weathering intensity. The most common clay mineral observed is illite. Smectite predominates in the cracking clay soils.

Kaolinite in the highly weathered soils of the intertropical zones (Southern India, South America, South-east Asia) and altapulgite or palygorskite (with gypsum and calcite) in the arid regions (Central and Southern Iraq, Western India).

The common mineral (in descending order) as observed in the dominant soil groups of India and Iraq are :

1. Alluvium – derived soils : illite (Vermiculite) Chlorite/ montmorillanite, Kaolinite (in flood plains and basin)

2. Forest Soils : illites, Chlorite (Vermiculite), montmorillanite, Kaolinite.

3. Black (Cotton) Soils : Smectite, illite, chlorite

4. Desert soils : illite, attapulgite (Smectite in basin depressions)

    Psendochlorite, Vermiculite (in normal cultivated soils)

5. Red & Lateritic soils : Kaolinite (gibbsite), chlorite, illite

Clay minerals – high surface area – more negative charge on them, they are a source of cation adsorption and cation release, considered important in basic soil fertility.

Non-Silicates

Oxides, Hydroxides or Hydrous Oxide Group of Minerals

Oxygen is present in great abundance (46.7%) in the earth's crust. The oxide minerals are formed by the direct combination of elements (present in earth's crust) with oxygen. e.g. $4Fe + 3O_2 \rightarrow 2Fe_2O_3$, $2Al + 3O_2 \rightarrow 2Al_2O_3$

Group :

The oxides are usually harder than any other mineral, except the silicates. The most important soil forming oxide minerals are

(1) Haematite Fe_2O_3, (2) Limonite Fe_2O_3 $3H_2O$ (3) Geothite FeO $(OH)_n H_2O$ (4) Gibbsite $Al_2O_3 H_2O$

1. Haematite Fe_2O_3

It varies in colour from red to blackish red and has a reddish streak. It has metallic luster and hardness about 5. Its presence in rocks is indicative of quick chemical change. Haematite alters to limonite, magnetite, pyrite and siderite. It occurs as a coating on the sand grains and acts as a cementing agent. It swells on absorbing water to form hydrated iron-oxide i.e. limonite $2Fe_2O_3$ $3H_2O$ and geothite FeO $(OH)_n H_2O$.

2. Limonite Fe_2O_3 $3H_2O$ or Bog Iron 2 Fe_2O_3 $3H_2O$

It is a hydrated ferric oxide, yellow to brown in colour and is of wide occurrence. It is the final product of most of iron minerals, and hence is resistant to any further change, except for absorption of water. It is an important colouring and cementing agent in soils. On hydration, limonite forms haematite. On reduction and carbonation, it yields soluble iron. Limonite is a common alteration product of pyrite, magnetite, hornblende and pyroxene. It may be present in the form of iron concentrations.

3. Geothite FeO $(OH)_n H_2O$

Usually white in colour but may be pink or grey in colour. Its hardness is 5.3

4. Gibbsite (Hydragillite) : $Al_2O_3 H_2O$

Aluminium compound in soils due to Gibbsite. Natural colour is white. It is abundantly found in highly weathered soils of the tropical zones, called Laterites (Oxisol). Its presence in soils suggests high degree of weathering and leaching under well drained conditions.

The red, yellow or brown colours in soils are due to the presence of geothite and haematite which occur as coatings on the surfaces of soil particles (especially clay).

2. Carbonate Group : The basic compounds like $Mg(OH)_2$ and $Ca(OH)_2$ combine with CO_2 or carbonic acid (H_2CO_3) to form carbonates as under, $Ca(OH)_2 + CO_2 = CaCO_3$ (Calcite) $+ H_2O$

$$Ca(OH)_2 + H_2CO_3 = CaCO_3 \text{ (*Calcite)} + 2H_2O$$

e.g. (1) Calcite $CaCO_3$ = A white mineral (hardness – 3) widely distributed in sedimentary rocks (like limestone) and decomposes easily as

$$CaCO_3 + CO_2 + H_2O \rightleftharpoons Ca(HCO_3)_w \text{ (Solubleform)}$$

(2) Dolomite – Ca Mg $(CO_3)_2$

Dolomite is less readily decomposed than Calcite; Dolomite is the chief source of Mg in soils.

(3) Siderite $FeCO_3$

It is an alternation product of other iron-bearing minerals (hardness - 4) and may it self alter to haematite or limonite. It is an important mineral in water logged soils.

3) Sulphate Group

Sulphate is a complex group formed by the combination of 1 sulphur & 4 oxygen ions, which further reacts with Ca to form Calcium sulphate (anhydrate $CaSO_4$). On hydration it forms gypsum ($CaSO_4$ $2H_2O$).

e.g. (1) Gypsum : Common mineral in desert soils and in sedimentary rocks with hardness of 2. It is slightly soluble in water and is most easily leached. It precipitates out as mycelium. From ground waters rich in $CaSO_4$ ions (as in the Mesopotamian plain of Iraq where aridic conditions prevail). In India, it is used as an amendment to reclaim sodic soils and acts as a source of Ca and S per plants. In Iraq its presence in high amount is a problem as it causes civil structures to collapse and make sink holes in soils, resulting in loss of scarce irrigation water.

4) Phosphate Group

Apatite : Rock Phosphate $(Ca_3(PO_4)_3)$

It is a primary source of phosphorus in soils. Its hardness is 5 in mho's scale. It decomposes readily under the influence of carbonic acid. It becomes immobile in soils as it readily combines with clays, Fe-Al hydrous oxides, Calcium carbonate etc. forming immobile constituents.

ROCKS : Their Formation, Nature and Classification

Petrology is the science of rocks

Pedology consists of (1) Petrography (2) Petrogenesis

(1)   Petrography – which deals with the description of rocks

(2)   Petrogenesis – which is the study of the origin of rocks

Definition

A rock may be defined as a hard mass of mineral matter comprising two or more rock forming minerals.

Originally the whole surface of the earth passed through a molten stage and the first solid rock was derived from this molten material known as magma.

Some rocks are hard and compact e.g. granite & basalt. While others are loosely aggregated, e.g. Conglomerate, sandstone. For identification certain characters such as structure, colour, sp. gravity, cleavage or fracture, and mineralological make up. These characters are not so well defined as in minerals.

Rock

Formation : Various process are involved in rock formation

1.   "Cooling and Consolidation of Magma"

Rocks are formed by cooling and consolidation of molten magma within or on the surface of the earth. e.g.Igneous or primary rocks.

2.   Transportation and Cementation of Fragmentary Material : Disintegration and decomposition lead to the breaking down of pre-existing rocks. The resulting fragmentary material is

either compacted insitu or transported in solution by the natural agencies of wind, water and ice to low lying areas, like oceans. Consolidation of these materials after their deposition results in the formation of rocks, called sedimentary or secondary rocks.

3. Alternation of Pre-existing Rocks : The primary and the secondary rocks when subjected to earth's movement and to high temperature and pressure and partially or wholly reconstituted or altered to new rocks, called metamorphic rocks.

According to the mode of formation, the rocks are divided into the following three main classes (Fig. 5) :

1. Igneous or primary rocks

2. Sedimentary or secondary rocks

3. Metamorphic rocks

1. Igneous Rocks (Latin ignis, means fire)

The igneous rocks are formed by the cooling and crystallization of molten material – magma – on or beneath the surface of earth. They are characterised by non-laminar-massive structure 95% of the earth's crust is with igneous rock. They are a source of parent material for other rocks and ultimately for soils.

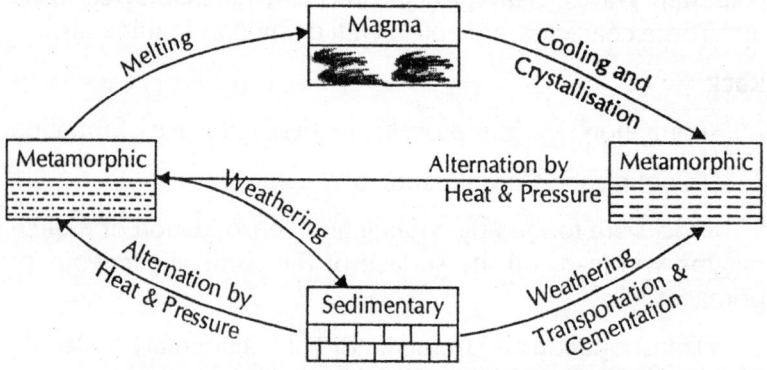

Fig. 5. A schematic diagram of rock cycle.

Formation : The molten mass or fluid lava is forced out to the surface of the earth, where it loses its volatile substances and cools down suddenly and solidifies forming crystals of fine size, the rocks thus formed have a glassy structure such rocks are called Extrusive or volcanic rocks.

e.g. Basalt, rhyolite, trachyte

The volcanic rocks may also be Vesicular, that is full of bubble holes formed by the escaping of volatile constituents, e.g. Pumice (Fig).

The rocks formed by the solidification of the magma within the earth's crust are called intrusive or plutonic rocks e.g. granite, gabbro.

Classification of Igneous rocks

	Kind of Rock → Kind of feldspar ↓	Chemical Composition (Classes)		
		Acid rock (over saturated) (> 60% of silica)	Neutral rock (Saturated)	Basic rock (Under saturated) (41-50% silica)
A.	Plutonic (Course grained)*			
	K-Na (Microclline)	Granite (10)	Syenite (15)	Foiidal syenite
	Na-Ca (Microcline)	Tonalite (20)	Diorite (20)	Gabbro (> 40)
B.	Volcanic (Fine grained)**			
	K-Na (Microcline)	Rhyolite (10)	Trachite (20)	Phonolite
	Na-Ca (Plagioclase Anorthite)	Dacite (20)	Andesite (40)	Basalt (> 40)

Fig. in paranthesis indicate the colour index
* Mineral grains can be seen by naked eye
** Mineral grains can be observed under a microscope or magnifying lens.

Based on Chemical Composition

Igneous may be divided into 3 (three) main classes depending on the relative amounts of acid and basic components

1. The greater bulk of the acid component is silicic acid or silica

2. Soda, potash, alumina, lime, magnesia and iron oxides constitute the main basis

1. > 60% silica – acid or over saturated rock e.g. granite

2. 41 to 50% + silica and large amounts of basic components termed as basic or under saturated rocks e.g. basalt

Weathering product of granite is light coloured – sandy soil with low specific gravity.

Basalt is dark coloured heavy soil with high specific gravity. e.g. Black cotton soils (Vertisols)

Brief Description of Important Igneous Rocks

1. Basalt : This is the most abundantly formed rock from molten material. It is a fine grained and dark coloured rock which cations 50% feldspar and 50% ferro magnesian minerals, including pyroxene, and olivine. The coarse-grained rock with the comparable composition is gabbro.

2. Pumice: It is a very light weight rock which has a lower specific gravity than water. As a result it floats on water. In composition, it is comparable to granite/rhyolite and in texture, it is like a sponge.

3. Granite : A coarse textured and light coloured rock that contains 60 to 70% feldspar, of which orthoclase constitute (40 to 45%), Plagioclase (20 to 25%), ferromagnesian minerals (3 to 10%) and quartz (20 to 30%), rhylite – the above granite composition.

2. Sedimentary Rocks

(Latin Sedimentum, means settling)

The sedimentary rocks are formed from sediments, derived from the breaking down of pre-existing rocks. The sediments are transported to new places and deposited in new arrangements and cemented to form secondary rock. Stratification is the most common feature in these rocks and as such these are also termed as stratified rocks.

Formation – 4 stages are recognized

(1) Weathering (2) Transportation (3) Deposition or Sedimentation (4) Diagensis.

1) Weathering : The igneous (primary) rocks and other rocks disintegrate due to physical chemical and biological weathering.

2) Transportation : The disintegrated material is transported by the agencies, such as water wind glaciers runoff and gravity. (The deposition of colluvium at the foot hills is due to gravity)

3) Deposition/Sedimentation : Disintegrated material or rock fragments are deposited through agents like water, wind, stream etc. This is how the coarser particles settle first and the finer particles later. This kind of deposition is called graded bedding when saturated with water, the loose material on gentle slopes may move, carrying with it the well established plants down the slope. This process is called solifluxion or soil creep.

4) Diagensis : Transformation of unconsolidated sediments to hard rock. It involves compaction and cementation.

(1) Compaction : The fine grained deposits under such environments are transformed to clays, shales etc.

(2) Cementation : The most common materials that serve as cementing agents, are lime, silica and iron oxide. Water that percolates carries the binding material/ materials in solution, deposit these in the voids of the loose sediments and bind the sediments together.

Classification : Based on their origin, the sedimentary rocks are grouped into 2 main classes :

1. Fragmental, Detrital or Mechanically Formed

These are formed by the deposition and cementation of erosion products (fragments) of pre-existing rocks e.g. Sandstone, Conglomerate, braccia and shale.

2. Chemically formed

(1) Inorganically formed rocks e.g. gypsum, halite (rock salt) (By evaporation) . These rocks are formed by the evaporation or precipitation of material dissolved in sea or lake water

Precipitation and floceulation are lime stone and dolomite (Calcareous rocks)

(2) Organically or biochemically formed rocks. By accumulation and partially decomposition of organic remains under anaerobic conditions. Depending upon the extent of change and the percent carbon retained, several varieties of carbonaceous rocks, such as peat, lignite, coal are formed

Brief description of Important Sedimentary Rocks

1. Conglomerate : Made up of more or less rounded fragments. If the fragments are more angular than rounded the rock is called braccia.

2. Sandstone : It is an intermediate between fine grained shale or mud stone and coarse grained conglomerate. The predominant mineral present is quartz. If quartz and feldspars, predominant are called arkose.

3. Shale : Fine grained rock composed ot clay and silt sized particles.

4. Lime stone : A rock, chiefly composed of calcite mineral.

3. Metamorphic rocks

The word metamorphic means – "Changes in form" and metamorphism issued as general term for all those changes.

Metamorphic rocks, therefore, are those which have undergone some chemical or physical changes from its original form.

In dynamo thermal metamorphism, the combination of pressure and heat forms more or less recrystallization of minerals with new structures. Typical rocks e.g. Schists and gneisses.

Pre-existing/Original rocks	Metamorphosed Rock
Conglomerate	Gneiss
Sand stone	Quartzite – quartz schist
Clayey sand stone (Impure)	Quartzite – Mica schist
Shale	Slate – Phyllite – mica schist
Lime stone	Marble
Dolomite (impure)	Dolomite marble – Soapstone or serpentine
Iron ores	Haematite schist
Basalt	schist (Chlorite schist, hornblende schist)
Gabbro	Gneiss – Chlorite schist
Granite & Syenite	Gneiss – Mica schist
Coal	Graphite

Classification : based on texture and structure of minerals, the metamorphic rocks are divided into (3) main groups, such as (1) Foliated (2) Unfoliated (3) Granular

Brief Description

(1) Gneiss : A crystalline rock, with banded appearance. The light coloured minerals such as feldspar, quartz & mica roughly alternate with bands of dark coloured ferro-magnesiuan minerals.

(2) Schist : A finely foliated or laminated rock, composed dominantly of micas and chlorite, together with some quartz and ferro-magnesian minerals. It can be derived from igneous and sedimentary rocks.

(3) Marble : A non-foliated crystalline rock, composed of calcite and dolomite.

(4) Slate – A very finely foliated rock with mineral particles too small to be seen. It splits into thin, smooth sheets. Mica, quartz and chlorite and are chief minerals in it.

Classification of metamorphic rocks (based on texture and structure)

Texture	Rock Name	Mineral Composition	Parent Rock
1. Foliated (Parallel structure) Latin-foliatus means – leaved or leafy The rock that contain micas and ferro-magnesian minerals and show foliation as the minerals are flattened and arranged in parallel layers.			
Coarse grained	Gneiss	Feldspar, quartz, micas, ampliboles,	Granite
Coarse grained	Schist	Micas and other elongated silicates with minor amounts of quartz (and feldspar)	Gabbrro basalt
Medium grained	Phyllite	Micaceons (Mica & chlorite) with larger grains representing a transition from schist to slate	Shale
2. Unfoliated (Massive structure)–The rocks containing quartz and feldspars do not show foliation even under pressure because of large sized crystals			
Fine grained	Anthracite	Carbon (92-93%)	Peat, lignite, coal
	Talc schist	-	Talc
	Ambibolite	-	Hornblende
3. Granular : The rocks consisting of mostly equidimentional grains			
Medium to coarse grained	Quartzite	Quartz & quartz cement	Sandstone
Fine grained	Marble	Calcite, dolomite	Dolomite calcite.

WEATHERING OF ROCKS AND MINERALS

Soils are formed from rocks through the intermediate stage of formation of regolith which is the resultant of weathering

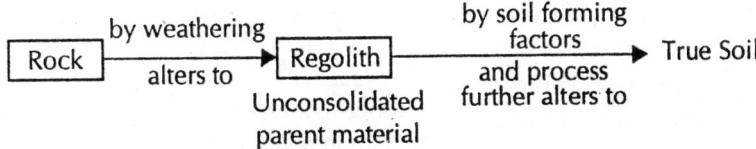

1. The formation of Regolith by the break down (weathering) of the bed rocks. (Weathering = The breaking down or disintegration of rocks by physical, chemical and biological process)

2. the addition of organic matter through the decomposition of plant and animal tissues, and reorganisation of these components by soil forming processes (eluviation and illuviation) to form soil which is three-dimensional natural body having unique characteristics (Fig. 6).

The initial stage of soil formation by the disintegration of bed rocks involves a set of processes, known by the term weathering. Weathering is, therefore, the process of disintegration and decomposition of rocks and minerals which are brought about by physical and chemical weathering, respectively, leading to the formation of regolith (unconsolidated residues of the weathering rock on the earth's surface or above the solid rocks). The regolith, or at least its upper portion, may therefore be termed as parent material of soils.

Parent material may be defined as the unconsolidated and more or less chemically weathered mineral material from which soils are developed.

The parent material are further acted upon by chemical and biological agencies which facilitates further grinding and leaching. Chemical weathering becomes more effective as the bed rock is broken down into smaller and smaller fragments because chemical reactions occur largely on the surface and therefore smaller the fragments, the greater is the surface area exposed to chemical reactions.

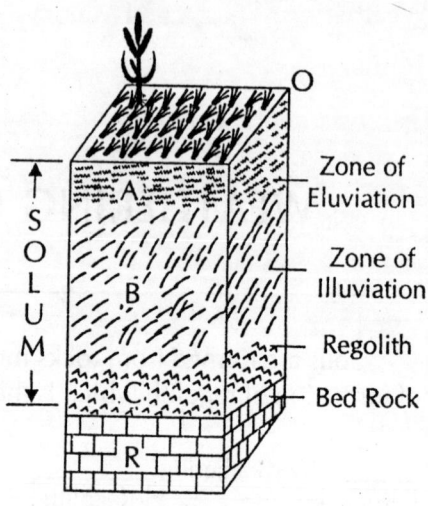

Fig (6). A typical soil profile - Horizon OABCR

Rocks, which are the original starting point in weathering process, are first broken down into smaller rocks and eventually into the individual minerals of which they are composed. Simultaneously the rocks disintegrate and the minerals therein are attacked by weathering forces and are changed to new minerals either by alterations or by decomposition, resulting in the release of soluble constitutents (most of which are lost in drainage waters) The minerals which are synthesized can be grouped as (1) the silicate clays and (2) the resistant end products, including iron and aluminium oxides.

Weathering of Rocks and Minerals
Terminology

Bed Rock	:	The solid rock underlying soils and the Regolith in depths ranging from zero (where exposed by erosion) to several hundred feet. or The solid rock at the surface of the earth or at some depth beneath the solum (soil)
Soil	:	a dynamic natural body composed of mineral and organic materials and living forms in which plants grow
Regolith	:	The disintegrated (and decomposed) mass of weathered rock and soil material overlying a solid rock on the earth's surface, in which the soil develops
Solum	:	The upper and most weathered part of the soil profile representing A and B horizons.
Decomposition	:	The breaking down of rocks into simpler compounds or elements by chemical and/or biochemical processes.
Eluviation	:	The process of removal of soil constituents from a horizon in solution or in suspension by the percolating water when there is an excess of rainfall over evaporation. The horizon that has lost materials through this process is called eluvial and the one which has gained material is called illuvial
Illuvial horizon	:	A horizon that receives material in solution or suspension form (by percolating water). The process is called Illuviation.
Parent material	:	The unconsolidated mass from which the solum develops by the soil forming factors and processes. The C. horizon is referred to as the parent material.
Disintegration	:	The process of breaking down of rocks into smaller pieces due to physical or mechanical processes
Weathering	:	The breaking down or disintegration of rocks by physical, chemical and biological process.

37

WEATHERING PROCESS

There are two basic processes of weathering viz., (1) Physical or mechanical and (2) Chemical

(1) Physical or mechanical = designated as disintegration and (2) Chemical = Decomposition

Another process which plays its role is biological disintegration

All these processes is a continuous chain of reactions

Table : Different weathering agents

	Physical/Mechanical (Disintegration)	Chemical (Decomposition)	Biological (Disintegration and decomposition)
1.	Physical condition of rocks	1) Hydration	1) Man and Animals
2.	Change in temperature	2) Hydrolysis	2) Higher plants and their roots
3.	Action of Water : Fragmentation and transportation	3) Solution	3) Micro-organisms
	Action of Freezing : Alternate wetting and drying Action of glaciers	4) Carbonation	
4.	Action of wind and sand blast	5) Oxidation and reduction	
5.	Atmospheric electric phenomena		

I. PHYSICAL WEATHERING

By physical weathering, the rocks are disintegrated, that is broken down into comparatively smaller pieces, without producing any new substance. The results of physical weathering can be compared with the hammering or crushing of the rocks under a great force. The different agencies that take part in the process of physical weathering are :

1. Physical condition of rocks

The permeability (the ease with which air, or plant roots penetrate into or pass through a specific horizon) of a rock is probably the most important single factor which determines the rate at which the rocks weather.

For instance, a coarse textured (porous) sand stone will weather more readily than a fine textured (almost solid) basalt.

Unconsolidated fine deposits of volcanic ash weather rather quickly as compared with unconsolidated coarse deposits, such as gravels, which may take much longer, because the water percolation between and not through the gravels.

2. Changes in Temperature

The variations of temperature, especially if quite wide and sudden, greatly influence the disintegration of rocks.

During the day, rocks get heated by the sun and expand. At night, the temperature falls and rocks get cooled and contract. This alternate expansion and contraction due to diurnal changes in temperature is more common in hot and desert regions where the surface of a rock weakens and even crumbles since the rocks do not conduct heat easily. Therefore, the temperature of a rock at its surface is very different from that of the part beneath. Moreover, the minerals within a rock also vary in their rate of expansion and contraction. The cubical expansion of quartz is twice as that of feldspars. The dark coloured rocks are subject to fast changes in temperature as compared with light coloured rocks.

This differential expansion of minerals in a rock surface generate stress between the heated surfaces and cooled unexpanded parts, resulting fragmentation of rocks. This process, with time, may cause the surface layer to peel off from the parent mass and the rock may ultimately disintegrate. This phenomena is called exfoliation.

Exfoliation : A weathering process in which thin layers of rock peel off from the surface. This is caused by alternate expansion and contraction of the rock surface due to heating (during day) and cooling (at night). This process is sometimes termed as "Onion skin – weathering".

3) Action of water

a) Disintegration, Transportation and Deposition :

Water acts as disintegrating, transporting and depositing agent. Water beats over the surface of rock as the rainfalls and starts its journey towards the ocean. The moving water has a great cutting and carrying force. It forms gullies, ravines and carries with it the suspended soil material of variable sizes (Gully – A shallow steep sided valley that may occur naturally or be formed by accelerated erosion – Gully erosion that forms gullies).

The transporting power of water varies. It is estimated that the transporting power of a stream varies as the 6th power of its velocity, that is the greater the speed of water, the more is its transporting power and carrying capacity. A current moving at a speed of 15 cm, 30 cm, 1.2 m and 9.0 m per second can carry fine sand, gravel, stones (of about 1 kg weight) and boulders of several tonnes respectively. The greater the amount of suspended matter, the quicker will be the disintegration of rocks. Hence the disintegration of rocks is greater near the source of a river than its mouth where it slows down steadily before meeting the sea.

A river carries an enormous load of sediments (silt and clay) in suspended form. This load when becomes too heavy for the moving water to carry, especially in plain areas where its speed is slowed down, is often shed resulting in rising of the river bed.

With time it may get filled up forcing the river to change its course (direction). The material deposited in the silted bed is the alluvium and the change in course of river is called meander.

Alluvium : A sediment deposited by the action of stream (s). It varies widely in particle size. The stones and boulders when present are usually rounded or sub rounded. It is soil forming material. Some of the most fertile soils are derived from alluvium of medium or fine textures, for instance, the Indo-Gangetic Plain Soils.

Plain Area : When a river is in flood, water overflows its banks and spreads over large areas on either side. This result in the formation of river levees and silting up of the basins and depression lands, leading to the formation of flat level plains. The materials (alluvia) deposited near the banks are coarser and those deposited for a way are finer in texture. Often a river deposits its finer silt load at its mouth, where it meets the sea and forms a delta. The finest particles, however, are carried further down to the sea. From such materials are formed marine deposits which are often mixed with shells and plants. In case a river flows into a lake and unloads its sediments into it, the material so deposited is termed as Lacustrine (Lacustrine material = Materials deposited by lake waters).

All these deposits have their reflections in the soils formed on them. That is why, we have been, till recently, using the terms alluvial soil, marine soil, lacustrine soil etc. for soils developed on such materials.

b) Action of Freezing

Frost is much more effective than heat in producing physical weathering. In cold regions, the water in the cracks and crevices freezes into ice and its volume increases by $1/10^{th}$ (approximately). The increasing volume due to freezing of subsurface water exerts an enormous outward pressure which breaks apart the rocks (e.g. keep the coca cola in freezer – There is breaking up of coca cola bottle).

c) Alternate Wetting and Drying

Some natural substance increases considerably in volume on wetting and shrink on drying e.g. smeetite clay (mantmorillonite) shrink and swell alternate wetting and drying – clay enriched rocks – disintegrate (platy structure) especially shake is to make them loose and eventually the rock breaks

d) Action of Glaciers :

In cold region, when snow falls it accumulates and some time turns into ice sheet. Ice is an erosive and transporting agent. On moving, the glaciers exert a tremendous pressure on rocks over which they pass. Glaciers ride over the big mountains, grind the rocks, crush the trees and/or buildings that come across their way. The loose material is pushed forward and carried by the moving glacier and is ultimately deposited on reaching warmer regions, where its movement stops with the melting of ice. The material left behind forms a structureless mass and is termed as Moraine. The huge boulders seen in the Kangra valley of Himachal Pradesh are examples of glacier movement and deposition.

(4) Action of wind and sand blast

Wind has both erosive and transporting effect when the wind is blown with fine material such as fine sand, silt and clay particles. It has a serious abrasive effect and the sand blown wind itch (hit) the rock, which ultimately breaks down under its force. Dust storm may transport tonnes of material from one place to another. This shifting of the soil cause serious wind erosion problem and may render the cultivated land as degraded for example sandy desert of Rajasthan and Southern Iraq.

(5) Atmospheric Electrical Phenomenon

This is also an important phenomenon during rainy season, when the lightening breaks up rocks and/or widens cracks.

2) CHEMICAL WEATHERING

In physical weathering rocks are broken down to finer

42

fragments. Chemical weathering is more complex in nature and involves the transformation of the original material into some new compounds (not present in the original rock) that is, it brings about alteration in materials.

for e.g. Chemical weathering of feldspar. Produces the clay mineral, which has a different composition and physical properties.

Feldspar + Water–Clay mineral + soluble cations and anions.

Chemical weathering becomes more effective as the surface area of the rock increases. Since the chemical reactions occur largely on the surface of rocks, therefore the smaller the fragments, the greater the surface area per unit volume of soil available for reactions. Chemical weathering is the most important process so far as soil formation is concerned. However the effectiveness of chemical weathering is closely related to the mineral composition of rocks. For instance, a mineral like Quartz (SiO_2) respond for slowly to chemical attack than does a mineral like olivine (Fe, $Mg)_2$ SiO_4 or Pyroxene.

Table : 1 Average mineralogical composition (%) of some important rocks

Rocks → / Mineral composition ↓	Granite	Basalt	Shale	Sand stone	Lime stone
Feldspar	52.4%	46.2%	30.0%	11.5%	-
Quartz	31.3%	-	2.3%	66.8%	-
Pyroxene amphibole	-	44.5%	-	-	-
Iron oxide minerals	2.0%	9.3%	10.5%	2.0%	-
Clay minerals (including mica)	14.3%	-	25.0%	6.6%	24.0%
Carbonates	-	-	5.7%	11.1%	76.0%

Chemical process they occur simultaneously in nature. Chemical process are (1) Solution (2) Hydration (3) Hydrolysis (4) Oxidation (5) Reduction (6) Carbonation (7) Integrated weathering process.

1)   Solution : Water is a universal solvent. Its solubilizing action is enhanced when it contains dissolved CO_2, organic and inorganic acids or salts in it. Most of the minerals are affected by solubilizing action of water, though by varying degrees. Some minerals such as halite (NaCl) dissolve readily in water whereas the solubility of some silicates such as quartz in water is very low. Solution helps in continuous removal of weathered materials but total removal by simple solubilizing action is very limited. In arid climates, due to paucity of water even water soluble minerals remain in rocks and sediments whereas these are completely washed away in the semi-arid and humid regions.

2)   Hydration : Hydration means chemical combination of water molecules with a mineral to form a new mineral. Many anhydrous minerals undergo hydration when they come in contact with water. Hydration reactions occur primarily on the surface and edges of mineral grains but may change the entire structure in simple salts. Some examples of hydration reaction are given in equation(1) and (2)

$$CaSO_4 + \tfrac{1}{2} H_2O \longrightarrow CaSO_4\, 0.5\, H_2O \qquad (1)$$
(Anhydrate)   (Hemihydrate)

$$CaSO_4 + 2 H_2O \longrightarrow CaSO_4\, 2\, H_2O \qquad (2)$$
(Anhydrate)   (Dihydrate)

Hydrogen is always accompanied by increase in volume. The characteristics of hydrated minerals are different from their anhydrated counterparts.

The hydrated minerals are usually soft and more readily weatherable. The absorbed water provides a bridge or entry way for hydronium (H_3O^+) ions or protons (H^+) to attack the structure. Cracking of certain rocks is due mainly to hydration of their mineral constituents. Under hot desicating conditions, dehydration (reverse of hydration) can also occur.

3) Hydrolysis : Hydrolysis is one of the most important processes in chemical weathering and results in complete disintegration or drastic modification (in structure and composition) of weatherable primary minerals.

Hydrolysis involves the partial dissociation of water into H^+ and OH^- ions. Pure water undergoes very limited dissociation, but in the presence of dissolved carbon dioxide, minerals and organic acid in it, the concentration of H^+ ion increases, resulting in an accelerated hydrolytic action of water. Hydrolysis is a double decomposition process and a hydroxide of some kind is usually formed. Water thus acts like a weak acid on silicate minerals as depicted in equation (3) and (4).

$$H_2O \longrightarrow H^+ + OH^- - (3)$$

$$KAlSi_3O_8 + H^+ \longrightarrow HAlSi_3O_8 + K^+ - (4)$$

(Orthoclase)

$$K(Si_3Al)Al_2O_{10}(OH)_2 + H^+ \longrightarrow (KH)(Si_3Al)\ Al_2O_{10}(OH)_2 + K\ (5)$$

(Muscovite)   Illite

The above example of hydrolysis reaction is replacement of interlayer potassium in micas by protons or hydronium ions to produce illite.

The products of hydrolysis are either wholly or partially removed by the percolating water, depending on the climatic conditions and permeability of the residual materials. They may also recombine with other constituents to form clays. In a way hydrolysis reactions may be considered as the forerunners (first stage) of the clay formation.

4) Oxidation : Oxidation is an important chemical reaction occurring in well aerated rock and soil materials where oxygen supply is high and biological demand is low. It is particularly important in rocks and minerals that contain iron, an element that is easily oxidized. In most of the primary minerals, iron is present in ferrous (Fe^{2+}) form. On oxidation, it undergoes the following reaction

$$Fe^{2+} - Fe^{3+} + e^- \quad (6)$$

Oxidation of iron is a disintegrative weathering process in minerals containing ferrous as part of their crystal structure. Reduction in size and increase in electrical charge on oxidation of Fe^{2+} to Fe^{3+} create electrical and structural imbalances in these minerals. Rocks containing ferromagnesian silicates such as pyroxenes, hornblende, biotite, chlorite and glauconite are susceptible to oxidation. In other cases, ferrous iron may be released from the mineral and is almost simultaneously oxidized to the ferric form. An example of this is the hydration of the mineral olivine and the release of ferrous ions from it which may be immediately oxidized to ferric form which has very low solubility. When ions such as Fe^{2+} are removed or are oxidized within the minerals, the rigidity of the mineral structure is weakened and the mechanical breakdown becomes easier. This provides a favourable environment for chemical reactions. Similarly minerals containing manganous form of manganese and sulphitde groups are also susceptible to oxidation.

5)   Reduction : Reduction occurs where a material is water saturated (such as below the water table) oxygen supply is low and the biological oxygen demand is high. The net effect of these conditions is to reduce a metal, say iron to its highly mobile ferrous form. In this form, it may be removed from the system if there is a downward and /or outward movement of ground water. If ferrous iron persists in the system, it tends to form sulphides and other ferrous compounds. These impart characteristic green and blue colours to many reduced soil materials. Under certain conditions, there could be formation of lepidocrocite (r-FeOOH), resulting in characteristic orange and yellow mottles.

6) Carbonation : Carbonic acid, although a weak acid of CO_2 is very important in chemical weathering of rocks and minerals as it makes minerals more soluble. The atmosphere contains only 0.33% CO_2 but the rain water may contain as high as 0.45% CO_2. The decomposition of organic matter liberates CO_2 in large amounts. Carbonation tends to produce carbonates and bicarbonates.

$$H_2O + CO_2 \longrightarrow H_2CO_3 - (7)$$

$$CaCO_3 + H_2CO_3 \longrightarrow Ca(HCO_3)_2 - (8)$$

(Calcite) (Carbonic acid)   Calcium bicarbonate

The solubility of calcium bicarbonate is considerably higher than that of calcite.

7) Integrated Weathering Processes : Different types of chemical reactions described above occur simultaneously and are interdependent. For example: Hydrolysis of a given primary mineral may release ferrous iron that is quickly oxidized to ferric form, which in turn, is hydrated to give a hydrous oxide of iron. Hydrolysis also may release soluble cations, silicic acid, and aluminium or iron compounds. These substances can recombine to form secondary silicate minerals such as silicate clays.

3) BIOLOGICAL WEATHERING

Unlike physical and chemical weathering, the biological or living agents are responsible for both decomposition and disintegration of rocks and minerals.

The biological life is controlled by the prevailing environments. For e.g. fungi are active in warm humid environment and bacteria in cool humid environments; the earthworms are scarce in acidic environments dominated by pine vegetation.

1. Man and Animal

The action of man in the disintegration of rocks is well known as he cuts rocks to build dams, channels, and construct roads and buildings. All these activities result in increasing the surface area of rock for attack by chemical agents and accelerate the process of rock decomposition. The role of animals is some what similar to man. A large no. of birds, animals, insects, worms etc., that live in rocks; by their activities they make holes in them and thus help in weathering.

In the tropical and sub tropical regions, ants and termites build galleries and passages, carry materials from lower to upper surface and excrete acids, such as formic acid. The oxygen and water, with many dissolved substances, reach every part of the rock through the cracks, holes and galleries, and thus bring about speedy disintegration.

Rabbits, by burrowing into the ground, destroy the soft rocks. Moles, ants and bodies of the dead animals, provide substances which react with minerals and help in decaying process.

The earthworms pass the soil through their elimentary canal and thus bring about chemical and physical changes in the soil material.

2. Higher Plants and their roots

The roots of trees and other plants penetrate into the joints and cracks of rocks. As they grow, the roots exert a great disruptive force and the hard rock may be broken apart e.g. pipal tree growing in wall which on growing result in the development of cracks in the wall. The grass roots may form a sponge like mass, conserve moisture and check erosion and help to form aggregates in soils. This helps in movement of air and water but find their way to the rock below for further action. Some roots go quite

deep into the soil and may open a sort of drainage channel. The roots running in cracks in limestone or marble in search of food produce acids which have a solvent action on carbonates and thus itch walls of fissures by dissolving a part of the rock with which they come in contact. The plants die, the leaves and other plant remains decompose and produce carbondioxide which is of great importance in weathering.

3. Micro-organisms like bacteria, fungi, actimonycetetes etc. play an important role in mineral decomposition and soil formation. The micro-organisms are closely associated with the decay of plants and animal remains and thus liberate nutrients for use of the next generation plants and also produce CO_2 and complex organic compounds which help in mineral decomposition.

FACTORS AFFECTING WEATHERING

Different minerals weather at different rates. Three (3) major factors which affect weathering of rocks and minerals are :

1) Climatic conditions

2) Physical and chemical characteristics of the rocks

3) Stability of minerals

1) Climatic conditions

The climatic conditions profoundly influence the rate and nature of weathering. Under arid conditions, the physical weathering predominates. The size of the particles decreases with relatively little change in the chemical composition of a mineral. The original primary minerals are prominent, whereas the content of secondary mineral is low. Physical changes due to temperature fluctuations and wind action are accompanied by only limited chemical changes. Consequently, the soils of arid regions are remarkably like the parent materials from which they are formed.

Similarly, in extremely cold climates, the rocks and minerals undergo mechanical disintegration with little modification in chemical composition.

In humid regions, the forces of weathering are more varied. In the humid tropical regions as in South India the year round high temperatures and the luxuriant plant growth provide optimum conditions for intensive weathering.

2. Physical characteristics

The physical characteristics that influence weathering include particle size, hardness and nature and degree of cementation. Rocks comprising minerals with large crystals, disintegrate easily than those with fine crystals because of pronounced expansion and contraction due to changes in temperature. However, once the rocks get disintegrated into smaller fragments, the finer crystals undergo relatively rapid chemical changes than the larger ones because of fine grained materials that provide greater surface area for chemical attack.

Rate of weathering also depends on the hardness and cementation. For e.g. a dense quartzite or sand stone cemented strongly by a slowly weatherable mineral can resist mechanical breakdown and presents only a small total surface area for chemical activity. On the other hand, porous rocks, such as volcanic ash, coarse limestone or limestone, are readily broken down into smaller particles and are easily decomposed.

3. Chemical and structural characteristics

Chemical composition and structural characteristics of minerals also influence the ease of their removal or breakdown. Some minerals such as gypsum ($CaSO_4$ $2H_2O$) or Calcite ($CaCO_3$) can be solubilized in water saturated with carbondioxide and can be easily removed from the parent material. Some minerals such as ferromagnesian silicates like olivine and biotite contain readily oxidizable ion (ferrous), their component ions are not very tightly packed in the mineral crystal structures, and can be easily weathered. In contrast, the relatively tighly packed nature of crystal units and lack of oxidizable iron in muscovite import considerable resistance to weathering.

4. Stability of Minerals

Minerals can be arranged in the order of stability or weatherability. In view of the differences in surface area and consequent reactivity, it is desirable to separate mineral particles into two classes

Clay size and sand silt size

Fig. Stability sequence of different minerals

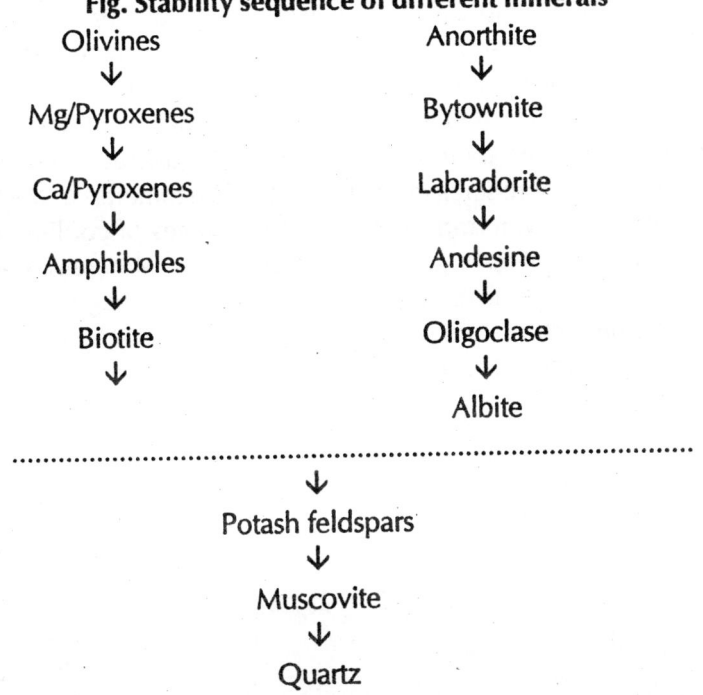

The weatherability of the common primary minerals is illustrated by the "Stability Series" (Fig.) proposed by Goldich (1938). On the left hand or the "basic" branch of this series, there is an increasing silica tetrahedral linkage with increasing stability from top to bottom. The least stable mineral (olivine) has independent tetrahedron and the silicon tetrahedral are held together by forming bonds with easily hydrozable magnesium or oxidisable iron.

On the other hand in Quartz, one of the most stable minerals, there is a three dimensional network of silicon tetrahedral. All the oxygen atoms are shared with adjacent tetrahedral. Also, there is a decrease in the content of easily hydrolyzable bases from the least to the most stable mineral.

On the right hand or feldspar branch there is a decreasing distortion of the lattice from calcic to potassic feldspars.

In Anorthite, 50% of the silicon in tetrahedral sites is replaced by aluminium whereas in potash feldspars, replacement is only 25%. Divalent calcium (in anorthite) does not fit well into the framework structure of feldspars. On the other hand, the large monovalent K fits appropriately in the feldspar structure. Thus, orthoclase is more stable than anorthite. The minerals as the top of both series were formed during early stages of cooling and crystallization of magma at high temperatures. Whereas quartz was the last mineral to crystallize from cooling of magma. Thus, higher the temperature of crystallization of a mineral from the magma, the more unstable it is.

6

SOIL FORMATION

At any specific location on the surface of earth, at least five factors are acting simultaneously to produce soil. These are (1) Parent material (2) Climate (3) Relief (4) Biosphere and (5) Time or age.

These factors are not of equal significance in development of different soils. Although some of them may be more effective in determining nature of soils under a particular set of conditions, all of them are inter-related and complement one another. Jenny (1941) expressed the relationship of these five factors to the soil properties by equation (9)

$$S = f (Cl, b, r, p, t......) - (9)$$

Where S = any soil property e.g. organic matter content of surface horizons, pH, soil, texture etc.

f = function of (or) dependent upon

Active {Cl = Climate

Active {b = biosphere (vegetation, organisms, man)

Passive {r = relief or topography

Passive{p = parent material

Passive {t = time or age

Thus, any soil property is a function of the collective effects of all the five soil forming factors

Joffee (1949) divided these factors of soil formation into two groups viz., active and passive factors of soil formation.

Passive factors represent the source of soil forming mass

and conditions affecting it. These are (1) Parent material (2) relief and (3) time.

Active factors represent the agents which supply energy that acts on the parent material for the development of soils. These factors are the driving forces that promote the processes causing changes in soil during the course of soil genesis.

Climate and biosphere are the active factors of soil formation. Soils are often defined in terms of these factors as "dynamic natural bodies having properties derived from combined effect of climate and biological activities as modified by topography acting on parent material over a period of time".

1. **Parent material :** Parent materials on which soils are developed can be divided into two broad groups, viz., Sedentary (formed in place) and transported. The transported material can be subdivided according to the agencies of transportation and deposition as shown in Fig.

Table : Transported parent materials.

Agent	Deposited by or in :		Name of Deposit or Parent material
WATER	(1) River (flood-plain terraces)	-	Alluvium.
	(2) Lake	-	Lacustrine
	(3) Ocean	-	Marine
WIND	Wind (aeolian sand dunes)	-	Dunes (of sand & Silt)
		-	Loess/Aeolian.
GRAVITY	Gravity action	-	Colluvium
ICE	Ice	-	Till Moraine

Different types of parent materials are described in the following sections.

1. Residual Parent Materials

When the soils develop at a place from the underlying rocks, they are said to have been formed from the residual parent materials, such materials develop in situ from the underlying rock. Typically, they have undergone prolonged and often intense weathering. In a warm, humid climate, as in some parts of Kerala, TN and Karnataka, the parent materials are likely to be thoroughly oxidized and well leached. Red and yellowish brown colours imparted by haematite ($Fe_2 O_3$) and geothite ($\alpha FeOOH$), respectively to minerals, are characteristics when weathering has been intense as in hot, humid climates. In cooler and especially drier climates, weathering of residual materials is less drastic and the oxidation and hydration of iron is hardly noticeable. On the other hand, such parent materials may be containing high calcium and magnesium due to leaching. Residual parent materials are encountered mainly in Deccan Plateau and some parts of central India while these are of relatively limited significance in the Aravalis and the Himalayan region.

2. Colluvial Parent Material

Rocks or soil debris at the foot of a slope that have moved there due to gravity are called colluvial materials. A colluvial material exists to some extent at the base of all slopes, but it is especially noticeable in mountainous or hilly topography where rock slides, slips are common. It is made up of fragments of rock detached from the higher parts and carried down the slopes mostly by gravity. The 'frost action' has also much to do with the development of such deposits. A parent material developed from colluvial accumulation is usually coarse and stony, because physical rather than chemical weathering is dominant. Soils developed from colluvial materials are not of significant agricultural importance generally, because of their small surface area, in accessibility and unfavourable physical and chemical characteristics. However, some useful timber and grazing lands in mountainous region have colluvial materials.

3. Parent materials transported by water

Sediments that are deposited by the flowing water such as

streams, rivers, etc. are called "alluvial materials". These deposits include flood plains, alluvial terraces, alluvial fans and deltas. These are described in the following paragraphs.

1. **Flood plains :** During rainy season, streams and rivers overflow their banks and flood the surrounding area. The parts of a valley and alluvial plains, which are covered during the floods, are referred to as Flood plains. Sediments carried by the swollen streams are deposited during the flood, with the coarser materials being laid down near the river channel and the finer materials further away. Such deposits are found to some extent along every stream. Depending upon the topography and lithology of the area, and the gradient of the river or stream, the flood plains may be very narrow or wide. At some places, the flood plain along the Brahmaputra and the Ganges are of 10 kilometer width. The soils derived from these sediments are generally rich in nutrients, but they may require drainage and protection from overflow during the rainy seasons. The flood plain soils are young and highly stratified.

2. **Alluvial Terraces :** The sedimentary deposits that were laid down by the rivers but are not subject to flooding now are called "alluvial terraces". These terraces are generally located at a higher elevation than the flood plains and are separated by escarpments (blocks) of different heights. Most of the Indo Gangetic Plains are alluvial terraces.

3. **Alluvial Fans :** Sediments deposited by streams of flowing water when they enter plains from the hilly areas are known as alluvial fans. Because of the presence of steep slopes, the flow velocity of streams in hilly and mountainous regions is high. When these streams leave a narrow valley in an upland area and suddenly descend a relatively flat and broader valley below, the sediments are deposited in the shape of a fan. Generally, these deposits are coarse textured, gravelly and stony, somewhat porous and well drained. Alluvial fans occur in the transition zones between the lower Himalayas and the Indo Gangetic Plains.

4. Delta deposits 5. Marine deposits 6. Lacustrine deposits 7. Glacial deposits 8. Glacial out wash sediments 9. Aeolian deposits

1. Parent materials and soil formation

The parent materials influence soil formation by their different rates of weathering, the levels of nutrients they contain and their particle size distribution. During the initial stages of soil development, the parent material exerts a greater influence in determining the properties of the soil. Thus, in slightly weathered soils such as those in flood plains, the parent material is dominant in determining soil properties. Soils developed on weakly cemented sandstones are sandy. Soils developed on shales are shallow and fine textured. Glacial till, loess or limestone deposits undergo very limited changes except accumulation of organic matter, development of some structure and loss of soluble salts.

The nature of parent material influences the soil characteristics to a great extent e.g. soil texture is highly influenced by parent materials. In turn, soil texture influences the down ward movement of water, and thereby affecting the translocation of fine soil particles and plant nutrients. The chemical and mineralogical compositions of parent materials also influence weathering directly and simultaneously, can affect the natural vegetation. For e.g. the presence of lime stone in the parent material delays the development of acidity, a process that humid climates encourage.

Different parent materials affect profile development and produce different soils, especially in the initial stages shows that under humid conditions

1. Acid igneous rocks (like granite rhyolite) produce light textured Podzolic soils (Alfisol)

2. Basic igneous rocks (basalt), alluvium or colluvium derived from limestone or basalt produce fine textured cracking clay soils (Vertisols) as in northern Iraq and Central India.

57

3. Basic alluvium or aeolian materials produce fine to coarse textured soils (Entisols or inceptisol)

2. Climate and Soil formation

Climate is a dominant factor in soil formation, mainly because of the effect of precipitation and temperature. Depending upon the quantum of precipitation in relation to temperature and evapotranspiration, we have arid, semi arid, subhumid and humid climatic regions. The dominant climates recognized are :

1. Arid climate : The precipitation here is far less than the water need. Hence soils remain dry for most of the time in a year.

2. Humid climate : The precipitation here is much more than the water need. The excess water results in leaching of salts and bases, followed by translocation of clay colloids.

3. Oceanic climate : Moderate seasonal variation of rainfall and temperature.

4. Mediterranean climate : The moderate precipitation, here is in winters and summers are dry and hot.

5. Continental climate : Warm summers and extremely cool or cold winters.

6. Temperature climate: Cold, humid conditions with warm summers.

7. Tropical and subtropical climate : Warm to hot, humid with isothermal conditions in the tropical zone.

Some direct effects of climate on soil formation include :

1. Retention or accumulation of lime (Carbonates) at shallow depths in areas having low rainfall. It is because calcium carbonate and bicarbonate (from dissolving carbon dioxide, minerals, and lime) are not leached out due to limited amount of water moving through the soil. Such soils are usually alkaline.

2. Formation of acidic soils in humid areas is due to intense

weathering and pronounced leaching of basic cations (Ca, Na, Mg, K).

3. Erosion of soils on sloping lands constantly removes developing soil layers.

4. Deposition of soil materials downslope buries the developing soils.

5. Weathering, leaching and erosion are more intense and a longer duration in warm and humid regions where the soil does not freeze. The reverse is true in cold climates, as in the central Himalayas. Water plays a key role in the soil formation. A soil is said to be developed when it has detectable layers (horizons), such as of accumulated clays, organic colloids, carbonates, or soluble salts that have been moved downward by water. The extent of colloid movement and the depth of their deposition are determined partly by the amount and pattern of precipitation, which produce the leaching action.

Climate also influences the soil formation indirectly through its action on vegetation. Semi-arid climate support scattered shrubs and grasses. Arid climates supply only enough moisture for sparse, short grasses, or shrubs which may not be dense enough to protect the soil against wind and water erosions. Many arid soils show very limited profile development and contain low amount of organic matter.

The soils formed under-the-year-round hot and humid climate are very deep, reddish in colour (due to the presence of oxidized iron as haematite) contain well decomposed organic matter and are low in essential elements because of intense leaching.

3. Biota and Soil formation (Biosphere)

The activities of living plants and animals, and the decomposition of their organic wastes and residues (the living environment, the biota) markedly influence the soil development. Differences in soils that have resulted primarily from the variability in vegetation are specially noticeable in the transition zone where trees and grasses meet. Under the humid forest vegetation, soils that develop may have many horizons but are leached (washed, eluviated) in the surface layers, and have slowly decomposing organic matter layers on the surface. In contrast, some grassland soils near the transition zone of forests have surface horizon rich in well decomposed organic matter, frequently extending down to a depth of 30 cm or more into the mineral soil.

Burrowing animals like rodents, earthworms, ants and termites are very important in the soil formation if present in large numbers. Soils that are habitat of many burrowing animals have fewer but deeper horizons because of the constant mixing within the profile, which nullifies the organic colloid and clay movements downward. Microorganisms help the soil development by decomposing organic matter slowly and forming weak acids that dissolve minerals faster than water. Some of the first plants to grow on weathering rocks are crustlike lichens, which are beneficial (symbiotic) combination of algae and fungi.

Buol et al (1973) recognized 4 major levels of organic horizons in mineral soil

Mor   : refers to surface soil horizon developed under acid litter and humus from coniferous vegetation, where fungi activity predominates.

Mull   : designated as forest soil horizon (Al) is of intimately mixed mineral matter and amorphous humus. It is slightly acid and is best developed under base rich litter, where bacterial activity predominates.

Sward   : is a dominantly rhizogenous Al horizon in grasslands as contrasted with zoogenous mull horizon of forest soils. This includes mollic epipedon, senso soil taxonomy, 1975, or Ap - horizon formed by cultivation of forest soils in general.

Orterde   : is a humus rich B horizon in podsols.

4. Topography and soil formation

The configuration of the land surface is known as topography or relief. Topography influences the soil formation primarily through its effects on modifying water and temperature relations. Soils within the same general climatic area developed the similar parent materials and on steep hillsides typically have thin horizons because limited amount of water moves down through the profile as a result of rapid surface runoff and also the surface erodes rapidly. Similar materials on gently slopping hillsides have more water passing vertically through them than do materials on steeper slopes. The soils on gentle slopes generally deeper, have more luxuriant vegetation, and the organic matter level is higher than in soils on similar materials on steep slopes.

The materials lying in land-locked depressions receive run off waters from surrounding higher areas. Such conditions favour better vegetation growth but exhibit slower decomposition of dead plants because of oxygen deficiency in water logged (saturated) soils. This results in soils with large amount of organic matter. If the area remains wet for many months of a year, organic (peat or muck) soils develop. If the accumulating waters dissolve salts from the surrounding soils, the depression may become a salt marsh with unique salt tolerant plants, or it may develop toxic salt conditions where no plants can grow. When soils in the water shed are strongly acidic, iron may leach from them and get deposited in depressions to form the bog iron (Limonite). Alkaline soils on the sloping topography in humid regions may result in lime leaching which gets eroded into depressions leading to these formation of marl (soft and unconsolidated $CaCO_3$ usually mixed with varying amounts of clay and impurities).

In the northern hemisphere, soils on south and west facing slopes receive more direct rays of the sun and are, therefore, warmer and drier than north and east facing slopes.

In arid climates, these drier south and west facing slopes are often less productive than soils on north and east facing slopes.

The opposite slopes are affected in the southern hemisphere.

In cold, wet areas these warmer sites may be highly productive. Higher temperature on south and west slopes results in greater loss of water by evaporation; the net result in regions where water is limited, is formation of soils with thinner horizons and less vegetative cover than the soils on north and east slopes.

5. Time and Soil Formation

The length of time required for a soil to develop the distinct layers called genetic horizons depends on many interrelated factors of climate, nature of parent material, organisms and topography. Horizons tend to develop most rapidly under the warm , humid and forested conditions when there is adequate water to move colloids. Acid sandy loams lying on sloping topography appear to be the soils most conductive for a rapid soil profile development.

Under ideal conditions, a recognizable soil profile may develop within 200 years. Under less favourable circumstances, the time may be extended to several thousand years.

Soil development proceeds at a rate determined by the combined effects of time and intensities of climate and biota (organisms) further modified by the effect of land relief (topography) on which the soil is situated and the kind of parent material from which it is developing.

Different surfaces of earth's lands have been exposed for different lengths of time. Some plateau soils have been exposed for hundred of thousand of years. Glacial till surfaces are more recent but may still be a few hundred thousand years old. More recently, rivers have flooded and covered flood plains and valley bottoms with recent deposition; these land surfaces may be only a few years or decades old, but the soil development is startling.

The soil on different aged surfaces have been forming of different lengths of time. Recent deposits show little soil development whereas land surfaces exposed for thousand of years may have well developed profiles that are quite different.

Minerals weather at different rates are classified, as per various schemes according to their weatherability. Mohr suggested (5) stages of weathering that are depended on mineralogical features of soils.

Sl.No.	Stage	Characteristics
1.	Initial	Unweathered parent material
2.	Juvenile	Weathering started; but much of the original material still unweathered
3.	Virile	Easily weatherable minerals fairly decomposed; clay content increased; slowly weatherable minerals still appreciable
4.	Senile	Decomposition reaches at a final stage; only most resistant minerals survive.
5.	Final	Soil development completed under prevailing environments

The soil properties also change with time

1. Nitrogen and organic matter content increase with time provided the soil temperature is not high (thermal, hyperthermal or megathermal).

2. $CaCO_3$ content may decrease or is even lost with time provided the climatic conditions are not arid. The soils from Punjab (Gurudaspur series) are non-calcareons because of decaleification.

3. In humid regions the H^+ concentrations increases with time because of chemical weathering

4. The horizonation (into A, B, C layers) develops with time

Since nature requires hundreds of years to develop a centimeter of soil, we should not allow our soils to erode and wait for nature to repeat its cycle.

Time and degree of maturity are the factors used in many systems of soil classification, for instance classification of soils into zonal, Intra zonal and Azonal soils.

In India, in the alluvial regions soils were classified on the basis of age of alluvium as Khadar(Young) and Bhangar (Old).

The factors which retard the soil profile development may be summed up as under.

1. Low rainfall

2. Severe wind or water erosion

3. Activity of burrowing animals

4. Constant accumulation and/or removal of soil material by deposition and/or weathering agencies e.g. silt deposition by irrigation water in the Mesopotamia.

5. Very steep slopes

6. High water table

7. Resistant parent material containing excessive siliceous minerals

8. High lime content in the parent material

SOIL PROFILE AND PEDOGENIC PROCESSES

Horizon differentiation and Development

A soil profile is a vertical section through a soil with two or more different layers or horizons extending from the surface down to the parent material. The mature soil profile shows a sequence of soil horizons or layers, essentially parallel to the soil surface and extending into the parent material. Each horizon has varying soil properties, such as colour, texture, structure, that are observed when a vertical section of soil is exposed. There are two main kinds of horizons: Organic (O) and mineral (A,B,C). The organic soils are rarely observed in the arid and semi-arid environment. A horizon is considered organic if its organic matter content is more than 30% and the mineral fraction has 50% or more clay, or the organic matter content is more than 20% and the mineral fraction has no clay. Wetness generally favours the increasing thickness of O-horizon and may qualify as peat when it is more than 40 cm deep and its organic matter content is more than 50%. The mineral horizons, on the other hand, are low in organic matter and overlie the rockmass. These have been designated as A, B or C depending upon their position in the profile, mode of formation, degree of weathering, colour, texture and structure etc (Fig.).

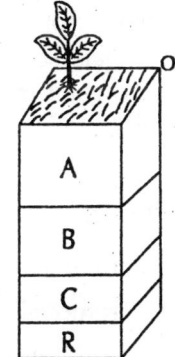

Fig. Soil Profile

PEDOGENIC PROCESSES

Pedogenic processes are extremely complex and dynamic involving many chemical and biological reactions, and usually operate simultaneously in a given area, one process may counteract another, or two different processes may work simultaneously to achieve the same result.

Fundamental Pedogenic Process

The basic processes involved in soil formation according to Simonson (1959) include

1. Gains or additions of water, mostly as rainfall, organic and mineral matter to the soil.

2. Losses of the above materials from the soil.

3. Transformation of mineral and organic substances within the soil.

4. Translocation or movement of soil materials from one point to another within the soil. It is usually divided into (i) movement of solution (leaching) and (ii) movement in suspension (elevation) of clay, organic matter and hydrous oxides.

All these 1,2,3 & 4 process promote horizon differentiation because of transfer of soluble salts, carbonates, silicates, clay minerals, sesquioxides and silica and transformation of minerals and organic matter resulting in the release of iron oxide and organic acids.

In contrast, the major changes that retard or offset horizon differentiation are due to

1. mixing of material by burrowing animals

2. removal of surface soil by erosion (water or wind)

3. creep (by shifting old and its replacement by new materials)

4. accretion of sediments in cultivated flood-plain areas, for instance silting of the irrigated areas in the Mesopotamia.

PEDOGENIC PROCESSES

Coming into being of a well formed soil from regolith or

newly deposited parent material is the outcome of several processes which are collectively called pedogenic or soil forming processes. Though physical and chemical weathering discussed earlier continues to operate even during pedogenesis, there are several other processes involving addition, leaching, redistribution, new synthesis or reorganization which lead to the development of a soil with distinct profile. Example of addition is the build up of humus or salts; that of redistribution is movement of clay or lime from the upper part of a soil profile to lower depths; that of leaching is removal from entire soil profile of soluble salts; and that of neo-synthesis is formation of organo-mineral or clay-humus or free sesqui-oxides. The development of soil structure is a reorganization phenomenon. Some of these processes such as redistribution of lime or build up of humus take place at a faster rate and reach a near equilibrium state in a few hundred to few thousand years, others like laterisation require a few thousand to millions of years to reach maturity.

An examination of a soil pit or a fresh road cut reveals distinctive horizontal layers. These layers will not be found if a similar cut were to be made in unconsolidated materials recently deposited by a volcanic eruption or flowing water during floods. Obviously, significant changes are made as soils develop from relatively unconsolidated materials. A study of soil formation (genesis) gives an indication of as to how these changes occur and why they can lead to formation of so many different kinds of soils. Soil genesis is brought about by a series of processes as mentioned above, the most significant of which are :

1. Humification

Humification is the process of transformation (i.e. decomposition) of new organic matter into humus. It is an extremely complex process involving various organisms such as bacteria, fungi, actinomycetes, earthworms and termites.

The waxy pine needles after falling on the ground are attacked by waves of fungi breaking down complex plant compounds, first the simple compounds, such as sugars and starches are attacked, followed by the proteins, cellulose and finally the very resistant compounds, such as tannins are decomposed and the dark coloured substance, known as humus, is formed. The whole process takes less than a decade.

In contrast, the decomposition of the leaves of deciduous trees by earthworms and bacteria is much faster and may be accomplished within a year or so, leaving little organic matter at the surface.

Micro-organisms control oxidation-reduction potential in soils. Under aerobic conditions, the liberation of CO_2, water and energy during microbial metabolism results in high oxidation potential of soil, normally known as biological oxidation. The result is in the loss of electrons, and lowering of pH in soils.

Under anaerobic conditions, there is gain of electrons and increase in soil pH, resulting in reduction of iron and manganese to form Fe^{++} and Mn^{++} which are toxic to plant growth. Besides waterlogging (anaerobic) conditions bring denitrification resulting in loss of free nitrogen. The yellowing of leaves (during monsoon) in areas subject to water logging is common feature, suggesting the effect of denitrification. Fertile soils have a higher oxidation reduction potential than the unfertile soils.

Nitrogen Transformation

The dominant forms of nitrogen transformations are ammonification, nitrification and nitrogen fixation. Ammonification – is the process whereby nitrogenous compounds in plant and animal tissues or decomposed to produce ammonium.

a) Weathering and organic matter break down, by which some constituents are modified or dissolved and others are synthesized.

b) Translocation of inorganic and organic materials up and down the soil profile, the materials being moved mainly by water but also by soil organisms and

c) Accumulation of soil materials in layers (horizons) approximately parallel to the soil surface.

The role of these major processes can be seen by following the changes that take place as soils form from relatively uniform parent material. When plants begin to grow and their residues are deposited on the surface of the parent materials, soil formation has truly begun.

Common soil forming processes are listed. Several of these processes operate simultaneously though some may start acting sequentially. An example of the latter is translocation of clay within a profile which occurs after leaching of lime or soluble salts. However, in the long run, one or two of these processes dominate and lead to the development of a soil with a distinctive profile dictated by factors of soil formation described earlier.

Some important pedogenic processes

1. Structure development – Grouping of individual particles (clay, silt and sand) together with humus and free sesquioxides into aggregates or peds of fairly distinctive size and shape.

2. Humification : Transformation of raw organic matter into humus.

3. Translocation of lime : Removal in solution of lime from the upper part of the profile and its partial or total accumulation in the lower part. The process leads to the formation of a Kankar or Calcic horizon.

4. Leaching : Removal in solution of a constituent from soil etc. soluble salts (to the movement and removal of materials in solution from the soil).

5. Salinization : Accumulation of soluble salts in soil.

6. Clay migration or Lessivage : Removal of clay particularly of fine clay in suspension from the upper part of soil profile

and its accumulation in the lower part. Eluviation : Washing out process of removal of constituents in suspension or solution by the percolating water from upper to lower layers. Dachanfour (1977) a French scientist used the term Lessivage for the mechanical movement of clay and iron oxides from the A-horizon without undergoing chemical alteration.

7. Braunification/rubification/ferruginisation : Release of iron from primary minerals and their disposal as coatings on soil particles or as complexes with organic matter/clay or as discrete aggregates to impart a brown to red colour to the soil.

8. Laterization : Removal of silica from soil and accumulation of sesquioxides (goethite, gibbsite, etc.) with or without the formation of iron, stone and concretions.

9. Podzolization : Removal of iron and aluminium, often as complexes with humus, from the upper part and its deposition at some depths.

10. Regur formation : Formation of intensely dark colour complex of smectitic clay and humus. This is a dominant process in black cotton soils.

11. Gleization : The reduction of iron under anaerobic (waterlogged) conditions with production of bluish to greenish grey colour with or without mottles or ferro-manganese concretions.

12. Illuviation – Process of deposition of soil materials (removed from the eluvial horizon) in the lower layer is termed as illuviation.

The Fundamental processes of soil formation which include transformation and translocation are

Specific pedogenic process.

13. Calcification : It is the process of precipitation and accumulation of ($CaCO_3$) calcium carbonate in some part

of the profile. Due to $CaCO_3$ results in the development of a calcic horizon. Calcium is readily soluble in acid, soil, water and when CO_2 concentration is high in root zone (Hydrolysis process). The depth of $CaCO_3$ enriched horizon depends on percolating rain water, ground water depth amount of rainfall and texture of soil. $CaCO_3$ enriched horizon are quite common in the arid or semi-arid environments (Rajasthan, Punjab, Haryana). Depth of occurrence : less than 50 cm in the arid 50 - 150 cms in the semi-arid, and more than 1.5 m in the sub humid region.

The lime enriched layer were termed by Marbut as Pedocals. Cal calcium to ped using vowel 'O'.

14. Decalcification : It is the reverse of calcification, i.e. that is the process of removal of calcium carbonate ($CaCO_3$) or calcium ion from the soil by leaching.

15. Podzolization : It is a process of soil formation resulting in the formation of podzols and podzolic soils. Podzolization is the negative of calcification. Podzolization leaches entire solum of $CaCO_3$ apart from Ca, the other basis are also removed whole soil becomes distinctly acidic.

16. Laterization : The term laterite derived from the word "later" meaning brick or tile and was originally applied to a group of high clay Indian Soil found in Malabar hills or Kerala, TN, Karnataka, and Maharashtra. It refers specifically to a particular cemented horizon in certain soils, which when dried, become very hard, like a brick. Such soils found (in tropics).

17. Gleization : The term "glei" is a Russian origin which means blue, grey or green clay. The process of soil formation in the development glei horizon (Hydromorphic soils – waterlogged soils) in the lower part of the soil profile above the parent material due to poor drainage condition (lack of oxygen) and waterlogged conditions prevail. Such soils are called hydromorphic soils.

Salinization is the process of accumulation of salts such as SO_4, Cl^{-1}, of Ca, Mg, Na, K, in soils in the form of salty (salic) horizon. As a result of the accumulation of salts, solonchaks or saline soil develop. Saline soil (EC of soln., > 4 dsm^{-1})

* Arid – semi arid climatic conditions

* Poor drainage/depressing land forms. Old later bottom. Alluvial deposits along the sea coasts/saline irrigation water

Desalinization : It is the removal, by leaching of excess soluble salts from horizon or soil profile by ponding water and improving drainage conditions by installing artificial drainage network.

Solonization/Alkalinization

The process involves the accumulation of Na ions on the exchange complex of the clay, resulting in the formation of sodic soils (Solonetz). Black alkali soils (Org. matter with Na)

Solodization or Dealkalization : The process of the removal of Na from the exchange sites. This process involves dispersion of clay. (hydration) Gypsum ($CaSO_4$ $2H_2O$) is added as an amendment mix before leaching. Final stage in the transition and resulting soils are known as soloth.

Pedoturbation : Another process which may be operative in soils is pedoturbation. It is the process of mixing of the soil. Mixing, to a certain content, takes place in all soils.

Faunal pedoturabation

Mixing of soil by animals such as ants, earthworms, moles, rodents and man himself.

Floral pedoturabation : Mixing of soil by plants as in tree tipping that forms pits and mounds

Argillo pedoturbation – Mixing of materials in the solum by the churning process caused by swell shrink clays as observed in deep black cotton soil of central India.

SOIL SURVEY

AIMS OF SOIL SURVEY AND TYPES OF SOIL SURVEY

Soil surveys began in 1895. They were simple at that time and were intended to answer practical agronomic questions about soil differences and limitations which were important for improving and expanding crop production. The soil surveys expanded in detail and concept with increase in the scientific knowledge and demand for more useful information.

In order to make optimum use of our limited soil resources, we need detailed information about their characteristics, type, and distribution on landscape. This information is collected through Soil Survey and Mapping.

Definition of Soil Survey : A soil survey is the systematic description of soils in the field, their grouping into well defined mapping units like soil series, phases etc. to identify their best use and show their location on map.

Aims (Objectives) of Soil Survey

The broad general objectives of soil survey could be grouped as fundamental and applied:

1. Fundamentally, a soil survey helps in gathering information about the properties, genesis, classification and nomenclature of soils.

2. The applied aspect in a soil survey includes interpretation of soil data for use in agriculture, forestry, pasture development, recreational purposes etc.

Both the fundamental as well as applied objectives are pursued with equal emphasis during a soil survey. Soil surveys are planned in such a way so as to provide information about soil to different users.

Uses of Soil Survey

Soil surveys are of great significance to any nation as they provide inventory of soil resources. Major uses of soil surveys are :

1. They provide information for the development of land use plans for both arable and non-arable lands and for predicting the long term effects of particular land use on environment.

2. They help in predicting the adaptability of identified soils to various uses and also their behaviour and productivity under defined sets of management practices.

3. Soil resource inventory helps in recognizing the areas having constraints like salinity, alkalinity, acidity, erosion, waterlogging, flooding etc. and in taking suitable measures for their management.

4. Soil information generated through soil surveys is useful for land settlement, tax appraisal, locating and designing highways, airports and other engineering structures.

In short, soil surveys provide information about the soils of a country and form the basis for land use planning.

Characteristics of soil survey

Soil survey are basically of two types

2. Single purpose soil surveys and

3. Standard soil surveys

1. Single purpose soil survey

These are designed for a specific objective with limited application such as soil fertility appraisal, crop suitability, soil conservation, soil erosion assessment, irrigation suitability, land settlement, trafficability, revenue and taxation etc.

2. Standard soil surveys.

These involve comprehensive data collection about soils and lands in such a manner that these could be used for a variety of purposes and encompass most of the single purpose soil surveys.

A soil survey involves a series of interlinked activities, some of which are

1. The study of important characteristics of soils and the associated external land features such as land form, natural vegetation, slope etc.

2. Laboratory analyses of soils to support and supplement the field observations.

3. Correlation and classification of soils according to the standard system of classification.

4. Mapping of soils, that is drawing and fixing soil boundaries of different kind of soils on a standard base map.

5. Transfer of agrotechnology through soil taxa which serves as a wheel for such transfers.

Practical Difficulties Encountered in Soil Survey

Surveyor's task is difficult and he is encountered with several practical constrains, such as –

1. The distribution of different soils may be so complex in a field or may occupy such small areas that their delineation on a map, at any practical scale, becomes difficult.

2. It is difficult to follow the geographic boundaries of soils because of the vegetative cover and their hidden nature.

3. Soil survey and mapping are expensive and many developing nations may not afford unless they serve the practical needs.

4. The inaccessibility of certain areas because of transportation problems may restrict the number of sampling points.

1. Base maps : Irrespective of the type of soil survey, a fundamental requirement of all mapping activities is a suitable base map. These base maps need to be complete in details of features and accurate in their location to enable the surveyor to delineate soil boundaries more correctly and conveniently. Depending on the intensity of mapping, following types of base maps are used for soil surveys in India and in many other countries.

   a) Cadastral Maps : Cadastral maps on the scale of 1:2640 (24" – 1 mile) to 1:7920 (8" – 1 mile) or 1:15,840 (4" – 1 mile) in plain areas and 1:1200 (52.8" – 1 mile) in hilly areas are used for detailed mapping. Cadastral maps show field boundaries and field or revenue survey number, however they lack the topographical details (contours, elevations etc).

   One advantage in using cadastral base is that the soil survey information or interpretations can be communicated to individual farmers by reference to the field survey number. The cadastral maps can be procured from the village 'Patwari' or concerned 'Tehsildar'.

   b) Topographical maps are published on the scale of 1:25,000, 1:50,000 and 1:250,000. These are used as base maps for soil surveys in India. Earlier, the scale of these maps used to be four inch to a mile (1:253, 400), one inch to a mile (1:63,360) or even two inch to a mile (1:31,680).

   Topographical maps show not only physical features but also contain topographical details in the form of contours and elevation above mean sea level. These maps have reliable planimetric accuracy facilitating measurement of distances and easy preparation of soil map. In India, topographical maps are prepared and published by the Survey of India, Dehradun and can be obtained from them or their regional officers.

c) Aerial photographs – are the pictures taken by camera fitted in an aircraft and flying over the terrain at a predetermined height depending on the scale of aerial photography and focal length of camera. Aerial photographs give a bird's eye view of large areas. The aerial photographs ranging in the scale from 1:8000 to 1:60,000 are used in different types of soil surveys.

d) Remote sensing : implies sensing from a distance. It includes aerial photography. Sensing devices located at a distance. It refers to the sensing or detection of electromagnetic radiation.

Kinds of Soil Surveys (Types of Soil Surveys)

Soil survey manual (Soil Survey Staff, 1951) and Soil Survey Manual of All India Soil and Land Use Surveys (AISLUS, 1971) recognized the following three types of surveys depending upon the objectives, methods, types of base material, intensity of field survey and type of map units.

1. **Detailed surveys**

2. **Reconnaissance surveys**

3. **Detailed Reconnaissance surveys.**

To these major types, semi detailed and exploratory or rapid reconnaissance surveys were added later, these surveys lead to small scale soil maps needed to macro level planning.

1. **Detailed surveys:** Detailed soil surveys furnish – information needed for the proper assessment of soil properties, terrain features, erosional aspects and related factors. Such surveys are time consuming, expensive and are recommended only for priority areas such as pilot projects, agricultural research stations, agricultural farms, micro-watersheds and for areas earmarked for urban development.

In the detailed soil map, the boundaries between mapping units are plotted on base maps (cadastral) or aerial photographs or satellite imagery from observations made throughout their course.

Cadastral maps on 1:2640, 1:7920 or 1:15840 scale and aerial photographs on 1:8,000 to 1:15,000 scale and satellite data on 1:12,500 scale are used, depending on the intensity of survey and the agricultural development needs of an area. In detailed soil surveys, the maximum distance between route or traverses is 250 m or even closer (50 - 100 m), depending upon the scale of the map and the complexity of the soil pattern.

2. **Reconnaissance surveys :** This type of survey is undertaken to prepare soil resource inventory of large areas. They furnish information to precede detailed or semidetailed soil surveys. On reconnaissance soil map, boundaries between the mapping units are plotted from soil observations made through auger sampling at an interval of 2.5 to 10 km. The representative profiles of various soils are studied at an interval of 3-6 km or even shorter, depending on the soil heterogeneity or variability. In reconnaissance survey, maps are prepared relatively at smaller scale, ranging from 1:250,000 to 1: 100,000 depending on the purpose. However field mapping is frequently done at 1:250,000 or 1:50,000 scale in order to use topographical maps, aerial photographs, or satellite imagery for the delineation of land types and planning of field traverses. At reconnaissance level, association of soil series, families and great groups are mapped. The variations in the land type, land use tone, texture, etc. as observed in aerial photographs or satellite imagery are used as the basis for boundary delineation.

2.(a) : Rapid Reconnaissance Survey

In this field mapping is done at 1:1,000,000 or still smaller scale using the satellite imagery. Soils are mapped by traversing representative areas. The mapping units are phases of great groups.

3.   Detailed Reconnaissance surveys

It is not really a separate kind (type) but in this type of survey, a part of the area is covered by detailed survey and the remaining by reconnaissance survey. It is carried out in individual areas best suited to each of these types. A reconnaissance survey is carried out over an area having limited potentialities for intensive development, in the usual way classifying and grouping the soils into soil series, or association of soil series as necessary followed by mapping. The remaining areas which show potentialities for priority attention are covered by detailed or high intensity surveys in the standard pattern.

3.(a) : Semi Detailed Survey – This comprises a highly detailed study of some selected sample strips cutting across many physiographic units and soils. This kind of survey provides sufficient information about various kinds of soil including problematic or degraded soils. Scale of base maps (aerial photographs or satellite imagery) used is 1:50,000. Mapping unit is association of soil series or families. The final maps are prepared on 1:50,000 scale.

9

REMOTE SENSING

Remote sensing implies sensing from a distance. It includes aerial photography, which is being widely used for soil mapping for many years now. Remote sensing is the science of obtaining information about objects or phenomenon in the environment through the use of sensing devices located at a distance without there being any contact between the object and sensing device. It refers to the sensing or detection of electromagnetic radiations which are reflected/scattered or emitted by an object. Earth orbiting satellites equipped with sensors, including cameras provide analog (imagery) and digital data. The sun's energy commonly referred to as electromagnetic spectrum (EMS) is an electromagnetic radiation that moves with constant velocity of light characterized by wave length or frequency. Non-photographic sensors can perceive the part of EMS from ultraviolet (wavelength less than 0.38 micrometre) through microwave to the upper wavelength of 100 cm. Remote sensing technology makes use of visible (0.4 – 0.7 μm), infrared (0.7 – 3.0 μm), thermal infrared (3 to 5 μm and 8 to 14 μm) and microwave (0.1 to 30 cm) regions of EMS to collect information about various objects on the earth's surface. Remote sensing technology provides accurate, timely and cost-effective information on resources.

Imaging from space has two important advantages; firstly, large area (thousands of square kilometers) can be examined from a single point in orbit; secondly, any area can be repeatedly examined on a regular basis for monitoring purposes. Because of the repetitive cycle of 16 to 26 days, enabling repeated collection of data for the same area, at the same local time, and the availability of satellite data on 1:25,000 to 1:250,000 scale, it can be used for semi detailed and reconnaissance survey of a district or a region for planning.

Satellite data from Indian, American and French satellites are being used for soil mapping by many national and state agencies in India. Before the launch of first Earth Resource Technology Satellite (ERST-1) in 1972 by the USA, aerial photographs were being used as a remote sensing tool for soil mapping. India has a well conceived earth resource satellite programme. Till date (January, 2002), Indian Space Research Organisation (ISRO) has launched seven Earth Observation Satellites starting with the launch of first Indian Remote Sensing Satellite - IRS –1A in 1988. The spatial resolution of the satellite data describe the minimum size of the object which can be separately identified and measured. The IRS-1C/1D-LISS-III data are available after every 24 days. One IRS-1C/1D-LISS-III scene covers an area of 141 km x 141 km.

Remote sensing data are a faithful record of ground details at the time of aerial photography or during recording by satellite sensor/camera. The satellite data have the same limitations as in aerial photography, but no differences in scale (due to tilt and changing height of the plane while flying).

10

LAND CAPABILITY CLASSIFICATION

LAND USE PLANNING

The land capability classification is the grouping of a land unit (s) into defined class (es) based on its capability. Land capability classification serves as a guide to assess suitability of the land for arable crops, grazing and forestry.

Some fields are suitable for cultivated crops, others for pastures, while still others may only be suitable for forestry or recreation. The use of land without paying regard to its capability is like withdrawing from bank account without knowing the balance. As a result crop yields progressively decline. It requires extra inputs and efforts to sustain crop productivity from such a declining soil resource.

The grouping of soils into capability classes and subclasses is done on the basis of their capability to produce crops and pasture biomass without adversely affecting the productivity over a long period of time. Klingebiel and Montgomery (1961) discussed the criteria for placing soils in different capability classes. The criteria is mainly based on (1) the inherent soil characteristics (2) the external land features (3) the environmental factors that limit land use.

The information on the first two items can be derived from the standard soil resource survey report and details about the third i.e. from environmental factors such as climatic and vegetation, can be collected from other agencies or sources. The factors which determine the capability of a soil are –

Depth of soil, stoniness, rockiness

Texture and structure of soil

Permeability (improvement of air and water through soil)

Relief, as expressed by slope

Extent of erosion

Susceptability to overflow and flooding and degrees of wetness.

Presence of toxic salts, alkali and other unfavourable chemical properties such as pH, gypsum, salt etc.

Severity of climate (temperature and moisture)

The land capability classification, however does not suggest the most profitable use of soils. Further these groupings are subject to change as new information about the soils and their response becomes available.

The capability classification consists of 3 categories

1. Capability – Classes 2. Capability subclasses 3. Capability units

1. Capability classes : In all (8) capability classes are recognised

Salient features of land capability classes

Land suitable for cultivation (I to IV)

Class – I (Green)

Very good cultivable, deep, nearly level, productive land with almost no limitation. Soils in this class are suited for a variety of crops, including wheat, barley, cotton, maize, tomato and beans. Need no special management practices for cultivation : Shown as green on maps.

Class-II (Yellow)

Good cultivable land on almost level plain or on gentle slopes that have slight limitation of soil depth salinity, texture, drainage or erosion that reduce the choice of plants. In general, these soils are suitable for wheat, barley, cotton, moderately suitable for maize, alfalfa, tomato and slightly unsuitable for beans. Recommendation is to cultivate with precaution, need simple management practices; shown as yellow on maps.

Class – III (brown)

Moderately good cultivable land on almost level plain or on moderate slope. These soils have limitation (s) of moderate erosion, soil depth, soil salinity, soil texture. They have vertic characteristics or drainage problem that reduces the choice of crop. In general, these areas have varying suitability for different crops. They are unsuitable for growing vegetable crops. Recommendation is to cultivate with careful management practices; need intensive care, shown as brown on maps.

Class – IV (pink)

Fairly good land on almost level plains or moderately steep slopes. Suitable for occasional or limited cultivation; generally unsuitable for growing a variety of crops because of strong or very strong soil salinity, shallow depth, erosion, fine texture or poor or excessive drainage. Suitable for selected crops and for pasture. Such soils may not be economical to cultivate as they need intensive soil conservation and management practices. Shown as pink on maps.

Land unsuitable for cultivation but suitable for permanent vegetation (Grazing)

Class-V (dark grey)

Land not suitable for arable farming, but very suitable for grazing; have limitations for use of implements due to stony or rocky and marshyness; shown as dark grey on maps.

Class – VI (Orange)

Non-arable land, well suited for grazing or forestry use. Have moderate limitations, such as steep slope, severe erosion, limited

84

soil depth, strongly gypsi ferrous, stony or sand dune areas. For instance, the dense forest lands of the Himalayas or gypsiferrous and dunal areas of SW Iraq shown as orange on maps.

Class – VII (Red)

Fairly well suited for grazing or forestry, not cultivable. Have severe limitations such as very steep land subjected to erosion or very shallow, stony soils having not enough available moisture for cultivation. Need careful management for grazing and forestry; shown as red on maps.

Class-VIII (Purple)

Non-arable, extremely rough, rocky, arid, wet or extremely saline land, suited only for wild life or recreation. Have very severe limitation, for instance highly eroded land, barren mountain tops (as in the Himalayas) or rocky undulating surfaces, shown as purple on maps.

2. **Capability subclasses** – Based on kinds of dominant limitations such as wetness or excess water (w), climate (c) soil (s) and erosion (e). The sub classes are mapped by adding limitation symbols to the capability class number subscriprs e.g. IIe, IIIw. Subclasses indicate both the degree and kind of limitations. There are no sub class(es) in capability class-I land, since there is no limitation in this class.

3. Capability Units

Further subdivisions of capability of sub classes. A capability unit include soils which are sufficiently uniform in their characteristics.

SOIL CLASSIFICATION

Pedological v/s edaphological approach

Classification is the grouping of objects in some orderly and logical manner. It is based on the properties of objects for the purpose of their identification and study.

Example : Soils are classified as sandy, loamy or clayey soil on the basis of their characteristics. For classifying the individuals of a large and widely varying population, such as soils, it is useful to group individuals into classes, and further into higher classes. This kind of grouping is called multi categoric or hierarchical system of classification. The individual soils are grouped into classes of lower category (e.g. soil series) which are further grouped into classes or higher categories (e.g. soil order). The lower categories are defined by a large number of differentiating characteristics and higher categories by a few differentiating characteristics. Within each class, there is a central care or nucleus to which the individual members are related in varying degrees. It is called the central concept or idealized individual which types of class.

I. Purpose of soil classification

Like the flora and fauna, soils are classified in a systematic manner, to remember their properties and understand their relationships.

The purpose of a classification

1. Organise knowledge leading to economy of thoughts

2. Recognise properties of the objects

3. Learn new relationships and principles in the population being classified.

4. Establish groups or subdivisions (classes) of the objects under study in a manner useful for practical and applied purposes in

   • Predicting their behaviour

   • Identifying their potential uses

   • Estimating their productivity

   • transferring agro-technology from research farms to cultivators' fields.

II. Evolution of soil classification system

Early systems :

The early systems of soil classification were quite simple and practical; their aim was utilitarian, for instance

1. Economic classification : It is the grouping of soils based on their productivity for the purpose of taxation.

2. Physical classification : It is the grouping of soils based on their texture a property of soil closely associated with soil productivity and management, for instance loamy, sandy and clayey soils.

3. Chemical classification : Grouping of soils based on the composition of soil having a bearing on their chemical characteristics, e.g. acidic, alkaline, calcareous, gypsiferrous soils etc.

4. Geological classification : Grouping of soils based on the nature of underlying parent rock/parent material e.g. basalt, limestone, sandstone etc. or transported material like alluvium, aeolian material.

5. Physiographic classification : Grouping of soils based on the characteristics of landscape, for instance levee, basin terrace, mountain valley, upland and lowland soil etc.

2. Dokuchaiev's Genetic System

In the later part of the 19th century, Dokuchaiev, working in Central Russian upland, observed that a uniform loess parent material extended for 100s of kilometers with increasing temperature gradient from north to south and an increasing rainfall and moisture gradient from east to west. These differences in climatic conditions were associated with important vegetation patterns, varying from forest to steppe (prairie) which left their important on the parent material, producing distinct soil differences. Such observations led Dokuchaiev the founder of the modern pedology to establish the concept of soil as an independent natural body and develop a series of publications on soil genesis and classification. The Russian approach tends to emphasize soil genesis, and hence the term genetic system of soil classification. Dokuchaiev (1900) divided soils into three categories. Normal, Transitional and Abnormal. These categories were later termed Zonal, Intrazonal and Azonal soils respectively.

Zonality concept : The soils that have fully developed soil profiles, and are in equilibrium with the environmental conditions, such as climate and vegetation are termed as zonal soils, for instance sierozem, chestnut, podzol and laterites. The soils formed in regions where time has been a limiting factor to produce fullydeveloped horizons are termed as Azonal soils, for instance, Alluvial soils and Regosols, lithosols. Still others occurring within the zonal areas and having characteristics that are determined largely by the local conditions, like topography, parent material, are termed as Intrazonal soils, for instance Calcimorphic and Hydromorphic soils.

Marbut's Morpho Genetic System

Morbut in the USA was greatly influenced by the Dokuchaiev's approach. He accepted the concept of the Russian soil type, but gave the name of Great Soil Groups. Marbut (1927) was the first to advocate classification of soils on the basis of their intrinsic properties rather than on soil forming factors, thus reducing emphasis on geology or parent rock. He published Atlas of American Agriculture. It was based on the iron alumina and lime contents. At the highest category level, he divided zonal soils into two classes: Pedalfers and Pedocals.

- Pedalfers : accumulation of iron & aluminium oxides (High RF areas – surplus of water leaching)

- Pedocals : Calcium as calcium carbonate (High evaporation – deficit of water)

Baldwin and Associates Genetic System

Morphogenetic system of Marbut (1935) was revised and elaborated by Baldwin et al (1938) and Kellogg & Thorp (1949). The system marked the beginning of a comprehensive approach. The salient features of the system are :

1. A return to the zonality concept of Russian School.

2. The pedocal pedalfer concept was deemphasized.

3. More emphasis was laid on soil as a three dimensional body with its own characteristics.

A new category viz., Soil family, was introduced between Great Soil Group and soil series; but neither soil families nor the higher categories were defined in terms of soil properties. The soils were grouped in three orders viz., Zonal, Intrazonal and Azonal following the Russians zonality concept, as under :

1. Zonal soils : The soils whose characteristics are determined primarily by the environment especially climate and vegetation.

2. Intrazonal soils : These soils occur within a zone, but reflect the influence of some local conditions, such as topography and/or parent material.

3. Azonal soils : The soils that have poorly developed profiles because of time as a limiting factor e.g. young soils without horizon differentiation.

The Three (3) orders were further subdivided into nine suborders on the basis of specific climatic and vegetative regions. Each suborder, in turn, was divided into Great Soil Groups which are expressions of more specific conditions. The Great Soil Groups were further subdivided into numerous soil families, series and soil types.

Limitations in the Genetic Systems

The major limitations in the genetic systems are :-

1. The 2 highest categories are defined in genetic terms and not on the basis of properties of the soils.

2. The concepts and definitions of the highest category, i.e. the order, in terms of soil properties, are not clear.

3. The Great Soil Group concepts and definitions are based on environmental factors, rather than on the soil properties. Hence their definitions are comparative and qualitative.

4. Many of the soils are defined in terms of properties with respect to soil conditions and are destroyed during the cultivation hence classification of such arable soil becomes ambiguous.

5. Very few properties accounting as a units in lower categories for interpretation.

6. The nomenclature in the highest categories laid too much importance on colour or vegetation rather than on the salient properties of the soils.

7. Nomenaculture was evolved from several languages. With mixture of nouns and adjectives, it was difficult to name the inter grades.

Hence a desirable system should be based on combinations of soil characteristics known to be significant to genesis and behaviour, but not directly on the either, In other words, the classification must be one that can be interpreted in terms of genesis and behaviour but the genesis and behaviour should be a step behind the classification.

12

SOIL TAXONOMY

A. Comprehensive system

In order to overcome the shortcomings of the genetic system, the US soil survey staff, in cooperation with many institutions, have been working since 1951 to arrive at a classification that narrows the differences in different view points. In 1960, a comprehensive system of soil classification popularly known as the "7th Approximation" was published. The system was put into official use in the USA and was adopted in many other countries, including India, Iraq, Belgium, The Netherlands etc. Ultimately in 1975, the system was brought out as Soil Taxonomy (Soil Survey Staff, 1975). Soil Taxonomy was designed to serve the needs of soil survey.

Salient Features :

Soil Taxonomy – Morphogenetic system in which morphology of soil, that is an outcome of soil genesis (Smith, 1963) serves as a guide.

Soil Taxonomy is based on the properties of soils as they exist today. Although one of its objectives is to group soils similar in genesis, the specific criteria used to place soils in different groups are those of soil properties. The system has an edge over the earlier systems in the following respects.

1. Unlike the Genetic systems, the comprehensive system is based on measurable soil properties that exist today.

2. It considers all such properties which affect the soil genesis or are the outcome of soil genesis.

3. The common definition of a class of taxonomic system is type or orthotype.

4. The nomenclature using coined words, is derived mainly from Greek and Latin language. Although it appears difficult once understood, it is the most logical nomenclature and helps in relating the place of taxon in the system and in making interpretations.

5. A new category i.e. subgroup, has been introduced to define the central concepts of great groups and their intergrades in order to express and recognize more clearly that soils are in continuum and show gradual change in many properties.

6. Unlike the genetic system, it is an orderly scheme without prejudices, but facilitates easy recognition of the objects.

Diagnostic Horizons

Diagnostic horizons are understood to reflect genetic horizons widely occurring in soils, which fairly well describe and define soil classes. Thus, a diagnostic horizon defined as one, formed through pedogenic processes and having distinct properties or features that can be described in terms of measurable soil properties. The diagnostic horizons are largely used not only for identifying soils but also in classifying them at various categoric levels, especially Great Groups. A number of diagnostic horizons have been defined in soil taxonomy. Based on their location in soil profiles, these horizons are of two types viz., surface and subsurface.

The diagnostic surface horizons are called epipedons (Greek epi, over, upon and pedon, soil) The epipedons are simply the uppermost soil horizons and include the upper part of the soil darkened by organic matter. They are not synonymous with a horizon.

Nine epipedons viz., folistic, histic, melanic, mollic, antropic, umbric, ochric, plaggen and grossarenic are recognized, but generally speaking, three of these, viz., mollic, ochric and umbric are of importance in India.

The Diagnostic sub surface horizons are called endopedons

(Greek endo dermis, sub surface or deep seated and pedon, soil). The endopedon includes the lower part of the soil where soil materials accumulate. Nineteen (19) endopedons viz., argillic, natric, agric, spodic, sambric, cambic, kandic, oxic, sulphuric, salic, placic, albic, glossic, calcic, gypsic, duripan, fragipan, petrocalcic, and petro gypsic, are recognized. Of these, eight (8) i.e. argillic, natric, cambic, kandic, oxic, salic, calcic, and gypsic are commonly observed in India.

Diagnostic surface horizons (Epipedons)

1. Folistic epipedon – surface layer, never saturated with water for more than 30 days (cumulative) in normal years (and is not artificially drained) (75% or more by volume of sphagnum fibre) BD (moist) less than 0.1 Mg m^{-3} or 15 cm or more thick, 16% or more organic carbon content if mineral fraction contains 60% or more clay. Folistic epipedons comprise organic soil materials that remain saturated for less than 1 month.

2. Histic epipedon : A thin organic horizon, very high organic matter content depending on clay content and that remains saturated with water for 30 days or more of the year. It is thinner than 30 cm if drained or 45 cm if not drained. The difference in Folistic and Histic in terms of saturation with water.

3. Melanic : A thick black horizon within 30 cm of the soil surface > 6% organic carbon > 4% organic carbon in all layers. The deep dark colour is due to the accumulation of organic matter resulting from root residues.

4. Mollic : A thick, dark coloured, soft mineral horizon with high (> 50%) base saturation and strong structure, > 1% org. matter. This epipedon is not naturally dry in all parts for more than 9 months in a year.

5. Anthropic – like Mollic, it forms large addition of org. matter(compost) > 1500 mg of citric acid (1%) soluble

P_2O_5 / kg of soil which distinguish it from mollic and umbric epipedon.

6. Umbric – like mollic but is low (<50%) in base saturation with high C:N ratio and is not naturally dry for more than 3 months in a year.

7. Ochric : Surface horizon i.e. light in colour. It contains <1% org. matter, hard, v. hard and massive when dry or remains dry for more than 3 months in a year.

8. Plaggen : A thick (>50 cm) man-made surface horizon, produced by a long and continued manuring. It can be easily identified by artifacts, such as bits of brick, pottery etc.

9. Grassarenic : A sandy horizon, > 100 cm thick over an argillic horizon.

Diagnostic subsurface horizons (Endopedons)

1. Argillic horizon – silicate clay enriched formed by illuviation of clay. Fine clay is deposited as clay skin or cutans on ped faces and on the walls of pores > 3% clay content (>40% clay (elutvial or bleached).

2. Natric : High Na-clay enriched, with columnar or prismatic structure. It is similar like argillic but >15% exchange complex saturated with Na or more exchange Mg + Na than Ca + exch. acidity at pH 8.2.

3. Agric : An illuvial horizon of clay silt and humus formed directly under the plough layer due to long and continued cultivation.

4. Spodic : A humus and/or sesquioxides enriched with or without iron. Formed in cold humid regions >85% spodic materials in a layer 2.5 cm or more thick. Such horizons are rarely observed in India. The so-called podsols – occur in the Himalayas do not meet the requirements of spodic horizon.

5. Sambric : A free draining, has colour (darkness) and base

status like umbric epipedon and has formed due to illuviation of humus and not of aluminium or sodium.

6. Cambic : Formed due to alteration by physical movement or chemical weathering, qualify it for argillic or spodic horizon. The horizon is extremely variable in mineralogy because of its pedogenic youthfulness. It may develop in the presence or absence of fluctuating ground water.

7. Kandic : With or without clay skins, CEC < 16 cmol (p⁺) kg⁻¹ of soil at pH 7 and effective CEC < 12 cmol (p⁺) kg soil. It shows no stratification, abrupt clear textural boundary.

8. Oxic : Enriched with Fe & Al oxides with dominance of 1:1 type of clay minerals and from where silica has leached 30 cm thick, sandy loam, kaolinite with CEC 16 or less c mol (p⁺) kg at pH 7.

9. Sulphuric : A mineral or organic horizon, pH of < 3.5 is toxic to plant roots and has yellow mottles of jarosite.

10. Salic : Secondary accumulation of water soluble salts (NaCl, Na_2SO_4 etc) at some depth in the soil profile. 15 cm thick, EC in saturated paste 30 dsm⁻¹/more the product of EC x thickness (cm) equal to 900 or more.

11. Albic : Bleached E horizon of podsols and planosols.

12. Glossic : (Gr. Glossa, tongue) - shows Albic characteristics gradually introducing into an argillic, a kandic, or natric horizon 5 cm or more thick and consists of an eluvial part which constitutes 15 to 85% (by vol.) of the glossic and an illuvial part.

13. Calcic : Secondary Ca & Mg carbonate enriched materials > 15 cm thick, > 15% secondary accumulation of carbonates and contains at least 5% more carbonates.

14. Gypsic : Ca or Mg sulphate enriched > 15 cm thick, > 5% $CaSO_4$.

15. Petrocalcic : An indurated calcic horizon hardness of 3 or more (Mhos scale).

16. Petrogypsic : A strongly cemented gypsic horizon whose dry fragments are not soluble in water.

17. Placic : Thin (2 to 10 mm thick) slowly permeable, dark reddish brown to black coloured iron or Mn pan that is lies within 50 cm of the surface.

Diagnostic Organic Materials

1. Fibric soil material (formerly peat) Fibric material over 2/3 of the mass.

2. Hemic soil material (formerly mucky peat or peaty muck). The fibrous materials 1/3 to 2/3 of the mass. An intermediate stage of decomposition.

3. Sapric soil material (formerly muck) – Fibrous material < than 1/3 of the mass.

4. Humilluvic material – Illuvial humus accumulate after prolonged cultivation of some acid organic soils.

5. Limnic material : Org. or inorg. materials deposited in water by action of aquatic – organisms or derived from under water and floating organisms ex. Marl, diatomaceons earth, and sedimentary peat.

Diagnostic characteristics for both Mineral and Organic Soils

1. Aquic conditions : Soils with aquic continuously or periodically saturated with water and undergo reduction.

2. Cryoturbation (Frost churning) : It is the mixing of soil matrix within a pedon that results in irregular or broken horizons, accumulation of O.M. over the permafrost table.

3. Densic contact (L. densus thick) : A thick contact between soil and densic materials below it has no cracks.

4. Densic materials : Relatively unaltered materials that do not meet the rquirements for any other named diagnostic horizons that have a non-cemented, rupture resistance class.

These are dense earthy materials. Does not allow root penetration.

5. Gelic materials : Mineral or organic soil material shows evidence of cryoturbation (frost churning). It is broken horizons, accumulation of O.M. generally on top of the permafrost and silt enriched layers. The structure is platey, blocky or granular.

Other Diagnostic Soil Characteristics

1. Abrupt Textural Change : boundary increase in the clay content within a short vertical distance between surface and sub surface.

2. Durinodes : These are weakly cemented indurated nodules, cemented by SiO_2.

3. Duripan : Sub surface horizon at least half cemented by SiO_2. The air dry peds are not soluble in water or HCl, but are destroyed by hot KOH after acid washing.

4. Fragipan : Sub soil layer of high BD. It is brittle when moist, and very hard when dry. It does not soften on wetting, but can be broken in the hands.

5. Low Chroma Mottles : These are moist soil colours of two or less chroma. 4 or more value often represent gley conditions.

6. Permafrost : It is a layer where soil temperature is always < 0°C. It may either be very hard or loose.

7. Plinthite(Greek, Plinthose – brick). It is a humus poor sesquioxide – rich horizon, which hardens irreversibly to ironstone, hard pans or aggregates with repeated wetting and drying. The red, indurating portions of the layer, are usually mottled with yellowish, greyish or white materials.

8. Soft Powdery Lime : It is authigenic lime translocated within the soil, normally present as coating on the ped surfaces.

9. Tonguing : It is used when albic horizon material (of at least 5 cm deep and 5 mm wide) penetrates into an underlying argillic or natric horizons.

Diagnostic contact (to "Non-soil" material)

1. Lithic contact : A boundary between soil and continuous coherent underlying material that has hardness of >3 on the Mho's scale and through which roots cannot penetrate.

2. Paralithic (Lithic like) contact: A boundary between soil and continuous non coherent underlying materials that has hardness >3 on Mho's scale. The roots can penetrate, at irregular and infrequent intervals, to 10 cm or more.

3. Petroferric contact : A boundary between soil and an indurated layer of iron cemented material.

Soil Moisture and Temperature Regimes

Control and guide soil utilization for plant growth. Classify the soil based on moisture.

Soil Moisture Regimes (SMR) refers to the presence or absence of water in a soil at different times of the year.

Soil is considered as Moist when moisture tension is less than 15 bar (1500 KPa) and Dry when the tension is 1500 KPa(15bar) or more within the Soil Moisture Control Section (SMCS). The limit SMCS by the soil depth to which the soil at wilting point is moistened when 2.5 cm and 7.5 cm of water are added at the surface. In general the upper and lower limits of SMCS in loamy soils are at 20 cm and 60 cm respectively.

SMCS remains moist with moisture tension between 33 KPa (1/3 bar) and 15 bar.

In classification of soil at different categories levels such as soil family, suborder e.g. order level Aridsol.

3. Dominant soil moisture regimes

1. Saturated   2. Non leaching   3. Leaching

1. Saturated – Taxonomically characterized as AQUIC when soil pores are completely filled with water resulting anaerobic conditions. It is not conducive to normal growth.

2. Non leaching : Taxonomically characterized as ARIDIC where water moves into SMCS for a very short period in a year and gets completely withdrawn by high potential evaporatranspirative demand (PET).

3. Leaching : Taxonomically characterized as UDIC where water moves into soil almost throughout the year (if not frozen).

As a thumb rule, any month with an average precipitation of more than 50 mm is considered to keep the SMCS partially moist.

Classification based soil moisture regimes

aquic	udic	ustic,xeric	aridic	torric
↓	↓	↓	↓	
gleying (saturated)	Moist soil throughout the year	Soils with limited moisture	Soils with negligible moisture	

Soil Temperature regimes play an important role in classifying soils at the family and suborder levels.

In all there are six (6) soil temperature regimes like (1) Pergelic (2) Cryic/frigid (3) mesic (4) thermic (5) hyperthermic (6) megathermic.

The prefix iso is sued if the difference between mean summer and mean winter temperature is less than 5°C to separate tropical areas.

Soil Temperature Regimes (STR)

Ranges of temperature within which biological activity of different degrees prevails :

Mean Annual Soil Temperature (MAST) (°C)

1. Pergelic   < 0°C Temp
2. a) Cryic   0 to 8°C   (Cold) soil
    b) Frigid   0 to 8°C   Cool to Soils (Warmer soil)
    c) Isofrigid   0 to 8°C
3. Mesic -   8 to < 15 °C   (Cool to warm soils)
    (i) isomesic
4. Thermic -   15 to < 22°C (Warm to hot soil)
    (i) isothermic   15 to < 22°C
5. Hyperthermic   22 to < 28°C (hot soil)
    (i) isohyperthermic 22 to < 28°C (Indian soils)
6. Megathermic   28° or more (very hot soils)
    (i) Isomegathermic 28° or more

The MAST can be computed from the mean annual air temperature (MAAT) by adding.

Definition of Soil Taxa

In the definitions of the taxa, differentiating characteristics selected are properties of the soils, including soil temperature and moisture regimes. The definitions are quantitative and precise rather than comparative and are written in "Operational" terms.

Nomenclature (Soil Taxonomy)

Most of the names used in the genetic system are biased in favour of colour (brown/red soils) and Vegetation (Prairie/forest soils) and show no similarity or relationship with each other. They also do not show the place of taxon in the system. In view of these limitations, the nomenclature used in Soil Taxonomy is altogether different and is based on coined words from Greek or Latin languages. The basic principles followed incoming the names, according to Heller (1963), are that the name should –

1. be most easily remembered

2. suggest some properties of the object

3. suggest the place of a taxon in the system

4. be as short as possible

5. be as euphonic as possible

6. fit readily in as many languages as possible

The names of the classification units are combination of Syllables. These names may sound strange at first, with experience, one can appreciate and use these with advantage. Each part of the name conveys a concept of soil character or genesis. For e.g. Aridisol (L aridus means dry, solum = soil) –the soils of dry places; Vertisol (L. verto = turn, solum = soil) – the soils which churn or invert.

Structure : The system has 6 (six) categories of classification (Fig.) from the highest to the lowest levels of generalization. They are grouped under two broad categories viz., higher and lower.

Categories (Soil Taxonomy)

↓	Order	12 order
Higher	Sub order	63
↓	Great Group	240
	Sub group →	1000
↓		
Lower	Family	
↓	Series	
↓	Series + Site factors	

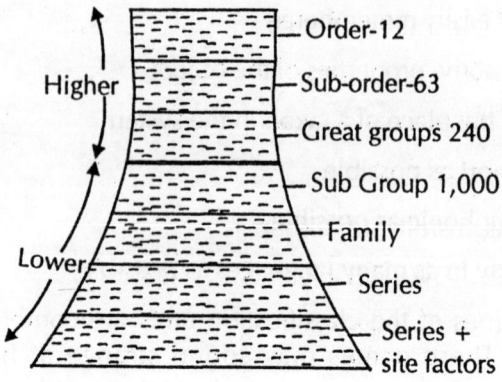

fig. The hierarchy of soil taxonomy

Higher categories

1. Order – The highest category in the system

2. Sub order – comparable to great soil groups of the genetic system

3. Great Group – A basic category, based on diagnostic sub surface horizons.

Lower categories

4. Subgroup – A new category designed to define the central concepts of great groups

5. Family – A practical category for making predictions for land use plans

6. Series – The lowest and the most specific category

These categories may be compared with those used for the classification of plants. Just as Acacia nilotica identifies a specific kind of plant, the Ghabdan clay loam or Tigris silt loam identifies a specific kind of soil. This similarity between the soil and plant classification schemes can be followed to the highest category, e.g. Phylum for plant and order for soil. The names of the orders are coined words have a common ending Sol (L. solum soil) with connecting vowel 'O' for Greek and 'i' for Latin and other roots

e.g. Mollisol – Order   OII→order

Ustoll – Sub order Suggestive property of the class Ustoll

Argiustoll – Great Group

Typic Argiustoll- Sub Group

Sub order – two syllables

1. Suggestive a property of the class

2. name of the order

Soil Order Phrases

AVAGAMI HOUSE   Memory device of Soil order

A = Aridisols – Arid – Desert, brown and reddish brown soil

V = Vertisols – Invert Black Cotton Soils (Regur) Grumosols

A = Alfisols – Pedalfer – Grey brown, non-calcic brown soils
                degraded

G = Gelisols – Frost churning – Frozen tundra soil. Cryoterbat

A = Andisols – Ando – Volcanic ash soils

M = Mollisols – Mollify – Chesnut Brown forest

I = Inceptisols – Inception – Ando, Forest Humic Gleysoil

H = Histosols – Histology – Bog soils

O = Oxisols – Oxide – Laterite soils, Lato sols

U = Ultisols – Ultimate – Red yellow podzolic half-bog soil

S = Spodosols – Podzol – odd – Podzols, Brown Ground Water–
                – Podzols

E = Entisols – Recent – A zonal soil Gley soil low, humic soil

Derivation of formative element

1. Entisol   -   Non sence syllable

2. Vertisol   -   L verto turn
3. Inceptisol   -   L. inceptum beginning
4. Aridisol   -   L. aridus, dry
5. Mollisol   -   L. Mollis soft
6. Spodosol   -   Gk. spodos, wood ash
7. Alfisol   -   Non sence syllable
8. Ultisol   -   L. ultimus last, Ultimum last
9. Oxisol   -   Frenchoxide, Oxide
10. Histosol   -   Gk. histos, tissue
11. Andisol   -   Jap. and
12. Gelisol   -   Gk. gel, ice

Key to Soil Orders

Key to Soil Taxonomy – To avoid confusion, one must use the key schematically represented in (fig.) to check differentiating characteristics for keying out soils

Concepts | Soil order |

Soils that have permafrost within   →Yes | Gelisols |
100 cm or gelic material with permanent
frost within 2 m of the surface

↓No

Soils that do not have > 30% organic matter →Yes | Histosols |
to a depth of 40 cm

↓No

Soils that have spodic horizon within 2 m but →Yes | Spodosols |
have no plaggen epipedon or argillic or Candic
horizon above the spodic horizon

↓No

Soils that have andic soil properties in 60%   →Yes | Andisols |
or more of the thickness between the soil

surface and 60 cm or lithic or paralithic contact

↓No

Highly weathered soils with an oxic horizon →Yes | Oxisols |
within 1. 5 m, but have no kandic horizon
or contain 40% clay in the surface 18 cm and
have a kandic horizon within 100 cm of the
surface overlying the oxic horizon

↓No

Soils with swell-shrink type clays, having 30% →Yes | Vertisols |
or more clay up to 50 cm depth or to lithic/
paralithic contact, deep cracks when dry, at
50 cm and have slicken sides or wedge-shaped
aggregates

↓No

Dry soils that have ochric or anthropic →Yes | Aridisols |
epipedon, and either have a salic or calcic,
gypsic, cambic, patrocalcic or dunpan or have
an argillic/natric horizon and an aridic soil moisture
regime and an epipedon that is not hard and
massive, when dry

↓No

Low base saturated soils that have an argillic →Yes | Ultisols |
or kandic horizon but with base saturation
(at pH 8.2) of < 35% at 2 m depth below
the surface.

↓No

Dark coloured base rich (> 50%) soils of →Yes | Mollisols |
grassland vegetation with a mollic epipedon
that is not hard and massive when dry

↓No

High base status (> 35%) soils of the humid →Yes | Alfisols |
and subhumid regions with an ochric epipedon

and an argillic/natric or a kandic horizon

↓No

Soils that have no spodic, argillic, natric,   →Yes │Inceptisols│
oxic, petrocalcic, plinthite, but have an
altered or cambic B horizon or an umbric,
mollic or plaggen epidedon

↓No

Recent soils with no diagnostic horizon   →Yes │ Entisols │
other than an ochric or anthropic epipedon

Fig. A flow diagram giving simplified key for classifying soils in different orders of soil taxonomy.

The goal of a classification system is to help soil survey in correlating soils and making predictions by assessing their potential and constraints and in agro technology transfer, using soil texa as a wheels for such transfer.

Land Use

1. Entisols : In the rocky humid or subhumid mountain regions medium textured, some time infertile (HP) or poorly drained for cultivation – land is used for raising a single crop in a year on conserved moisture.

2. Inceptisols : These soils occur almost throughout the world. In India these are observed all over except in hot and arid regions.

   Land use – Agriculturally productive and provide excellent natural grazing grounds.

3. Vertisols : Because of their shrink-swell character and development of deep, wide cracks, silken sides, these soils pose many tillage problems during cultivation.

4. Mollisols : Inherently the best agricultural soils of the world. In India, these soils produce optimum yields both under

irrigated and unirrigated conditions with minimum inputs.

5. Aridisols : These soils do not support a crop without irrigation, if irrigated for cultivation they pose the risk of salinisation.

6. Alfisols : In India the red soils are predominantly observed fertile and productive forestry, grazing, fruit, vegetables and grain crops. In India they are used for growing crops like wheat, maize, sorghum, rice, horticulture crops, mango, cashew and jack fruit.

7. Spodosols : These are mineral soils with accumulation of sesquioxides and humus in the subsurface horizons. These develop under cool, humid climate and coarse textured siliceous parent material which favours free leaching conditions. In India it is rare because lack of typical parent material/environmental conditions.

8. Ultisols : Low fertility, low base status these soils pose limitations for agricultural use. Generally, these soils are used for forestry. In the tropical regions soils are cultivated for pineapple, sugarcane, coffee, cocoa, coconut, rubber etc.,

9. Oxisols : One chemically degraded soils and need careful management for agricultural use. These are mainly used for grazing and forestry. Shifting agriculture is a common practice adopted in such areas.

10. Histosols : Best left under natural vegetation located in depressions and flat areas – peat, muck and bog soils

11. Andisols : Volcanic ash material, dark coloured, low BD soils, occur in steep slopes soils with high phosphorous fixation capacity P. is unavailable.

12. Gelisols (gelic a frost churning) : Most recently introduced one soil with permafrost occurring in the extreme northern hemisphere. Such soils do not occur in India.

13

SOILS OF INDIA

India, situated between the latitudes of 08°04' and 37°06'N and longitudes of 68°07' to 97°25'E, has a geographical area of 329 Mha. Physiographically it can be divided into the three broad regions, viz. Peninsula (a triangular Plateau in the Deccan and south of the Vindhyas), Extra-Peninsula (mountain region of the Himalayas), and the Indo-Gangetic Plain separating the two above mentioned regions.

Geologically, a great part of the Peninsula is occupied by Archean rocks comprising gneiss, schists and other rocks of diverse nature. Red soils (*Alfisols*) generally predominate in this region. Next in order of age are the *Cuddapah* and *Vindhyan* rocks, followed by the coal-bearing Gondwana formations supporting rocks of Mesozoic and Tertiary groups. These are mainly distributed over the north-central and NE-central parts where Red soils (Alfisols, Ineceptisols and Entisols) have developed. The western and central parts are covered by lava flows of the Deccan Trap where basaltic rocks predominate. Here, Black cotton soils of different thickness (Vertisols, Inceptisols) predominate.

The Extra-Peninsula, on the other hand, shows the development of marine sediments of all ages, especially in north of the Himalayas. The major rock formations are Tertiary - old sedimentary (sandstone, limestone, etc) and igneous (granites) (at places metamorphosed to gneisses and schists). Here, non-calcic brown soils (Inceptisols, Alfisls, Entisols, Mollisols) predominate.

The vast Indo-Gangetic and other plains of Pleistocene origin are composed of alluvium of the great river systems flowing in this region. These are the alluvial soils. However, depending on the age of alluvia and degree of development, they can be classified in the Orders, Inceptisols, Entisols or Alfisols in Soil Taxonomy.

Climatically, India shows three distinct seasons, viz. cool and mainly dry winters from November to February, hot and mainly dry summers from March to June and monsoon rainy season from mid-June to September. Rainfall is received during June to September (from the south-west monsoon) and during December to February (from the north-eastern winds). The Western Ghats and the eastern Himalayan ranges receive the maximum rainfall because of their alignment across the summer monsoon winds. The Indo-Gangetic Plains receive moderate rainfall; Rajasthan receives little rainfall as the Aravallis lie along the path of moisture-bearing winds.

The other climatic element, viz. temperature, has particular significance for plant growth, especially in northern India during winter; high temperature promotes evapotranspiration during summer and causes crop-affecting aridity. Peak temperatures of 42 to 47°C are observed during summer (May to June), which fall sharply with the onset of monsoon rains (in July to September). Sub-zero temperatures are observed during winters in the extreme northern regions where germination of seed becomes a problem.

The bio-climate is the climate prevailing a few metres above and below the soil surface where biological activity prevails. The different climatic elements of relevance to plant growth are expressed in the bioclimatic map (Figure). The soil temperature and moisture regimes as discussed earlier, play an important role in classifying soils and crop planning as each plant species requires, specific soil temperature and moisture conditions for its optimal growth. India is dominantly represented by thermic and Hyper-thermic temperature and Ustic, Aridic and Udic moisture regimes.

India with a great variety of landforms, geological formations and climatic conditions, exhibits a large variety of soils; the variety is so diverse that barring a few soil orders (Andisols, Spodosols), India represents all the major soils of the world.

The grouping of these soils can be achieved by using either of the two systems, viz. Genetic and Soil Taxonomy. Whereas the Genetic System is based on genetic factors and processes, the Soil Taxonomy is based on the properties of soils, which are the reflection of soil genesis and are measurable. The major soils of India, according to the Genetic approach, can be classified into a few soil groups, viz. Alluvial, Black, Red, Forest and Desert soils. But in terms of Soil Taxonomy, they key out in ten out of the twelve Soil Orders and a large number of great groups. Since the objective of this section is to apprise students of the major soils of India, it is considered desirable to do so in terms of a few soil groups as per the Genetic approach rather than explaining a large number of soil groups in terms of Soil Taxonomy. For instance, the Alluvial Soils (in Genetic System) can be keyed out in three or more soil orders, viz. Entisols, Inceptisols, Alfisols, or Ultisols (in terms of Soil Taxonomy) depending on the age and degree of development. Hence, the major soils of India are discussed in genetic terms giving their equivalents in Soil Taxonomy (Figure).

Alluvial Soils

The name Alluvial is given to the soils that have developed on alluvium, irrespective of their place of occurrence and degree of development. They are one of the important groups of soils for agricultural production. They are extensively distributed in the states of Punjab, Haryana, Uttar Pradesh, Uttaranchal, Bihar, West Bengal, Assam and coastal regions of India and occupy an estimated area of 75 Mha in the Indo-Gangetic plains and Brahmaputra Valley alone.

The parent material of these soils (alluvium) is of recent origin and has been derived from the deposition of erosion products brought and laid down by various river systems; coastal alluvium is however laid down by the sea currents.

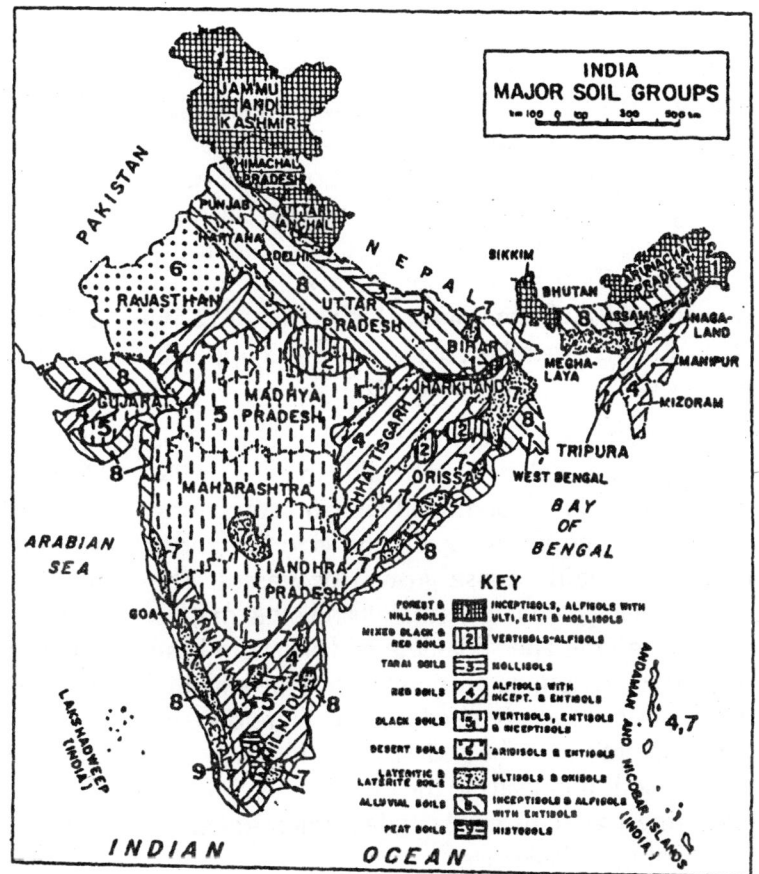

Fig. Major soils of India (Source : NBSS&LUP, 1985)

The major constrains of some of these soils are stratification that restricts leaching and drainage, extreme sandy nature that promotes excessive leaching of water and plant nutrients, hydromorphic condition that promotes reduction and results in poor aeration for plant growth. These soils, if managed well, can be fruitfully used for most of the agricultural and vegetable crops. They can be rendered saline where evaporation exceeds rainfall (aridic zone).

Black (Cotton) Soils

The name black is given to soils that are very dark in colour and turn extremely hard on drying and sticky and plastic on wetting, and hence are very difficult to cultivate and manage. These soils are comparable with the Grumosols of the USA. In view of their black colour, they are comparable with the Chernozems of Russia and Prairie soils of the USA, but differ in their physiochemical properties. These soils are dominantly distributed, in the central, western and southern states of India (Fig.). According to recent study, different members of black soils occupy an estimated area of 74 Mha.

Typical black (Cotton) soils, with characteristic swell-shrink nature, have developed on basaltic material (rock or alluvium) under semi-arid to sub-humid climatic conditions. The natural vegetation of the area is tropical dry deciduous and tropical thorny forests with fairly dense growth of grasses. The weathering product of the parent rocks (basalt and other metamorphic rocks, containing lime and soda-lime feldspars) are basic and rich in clay, which is dominantly of smectitic type with high co-efficient of expansion and contraction, and therefore set up a steady churning process in the pedon. Churning causes vertical mixing in deep soils and leads to the development of deep (> 50 cm) and wide (> 1cm) cracks, gilgai micro-relief and/or closely intersecting slicken sides.

The major constraints in their land use are: narrowing of workable moisture range, low infiltration rate and poor drainage, poor moisture and nutrient availability for plant growth, poor in some available plant nutrients, especially nitrogen, phosphorus, sulphur, their calcareous nature adversely affects the availability of micronutrients, and due to swell-shrink nature these are unsuitable for laying foundations, construction of buildings, laying of pipelines and electric communication poles etc.

The soils are inherently very fertile and, under rainfed conditions, they are used for growing cotton, sorghum, millet, soybean, pigeon pea, etc. Under irrigated conditions, they can be used for a variety of other crops, such as sugarcane, wheat and citrus plantation.

Red Soils

The name Red is given to soils rich in sesquioxides that have developed on rocks of Archean origin (granite, gneiss) and on well drained, stable and higher land forms under hot, semi-arid to humid, substropical climatic conditions. Under such conditions, the weathering is moderately intense and leads to enhanced decalcification. Some weathering products are leached out leaving behind the less mobile elements, like silica, iron and alumina. The iron and aluminium under oxidized conditions, form sesquioxides (Fe- and Al-oxides), imparting red colour to these soils.

These soils are predominantly observed in the southern parts of the Peninsula comprising the states of Andhra Pradesh, Tamil Nadu, Karnataka, Maharashtra, Orissa and Goa and in NE States. In Andhra Pradesh, the Red and the Black soils occur under similar bioclimatic conditions, but on different parent material and landforms. The Red soils develop on igneous (acidic) rocks and occupy higher topographic positions whereas the Black soils develop on basalt (basic) rock or on alluvium derived from basalt, and occupy lower positions on the landscape. The soils grade from shallow, gravelly and light-coloured (of the uplands) to much fertile, deep, dark reddish brown in the plains and valleys.

The soils pose limitations of soil depth (on hills and hill slopes), poor water and nutrient holding capacity, surface crusting and hardening, excessive drainage and runoff, poor natural soil fertility (N, P, Ca, Zn, S). Under good management practices, these soils can be profitably used for a variety of agricultural horticultural and plantation crops (depending on the moisture regime) such as millets, rice (both direct seeded and transplanted), groundnut, maize, soybean, pigeon pea, green gram, jute, tea, cashew, cocoa, grapes, banana, papaya, mango, etc.

113

Laterite and Lateritic Soils

The term 'Laterite' was originally used by Buchanan in 1807 for the highly ferruginous, particular and unstratified material observed in Malabar hills of south India. The laterites are specifically formed in tropical climate experiencing alternate wet and dry seasons. With monsoon type of climatic conditions acting on the basic parent rock, the siliceous matter is leached almost completely during weathering and the sesquioxides are left behind. On drying, these are converted into irreversible iron and aluminium oxides. The soils thus formed are rich in sesquioxides, devoid of bases and primary silicate minerals, hard or capable of hardening like bricks when exposed to drying after wetting. It is a compact to vesicular rock like material composed of a mixture of hydrated oxides of iron and aluminium with small amounts of manganese oxides and titania. They are generally observed on hill-tops and plateau landforms of Orissa, Kerala, Tamil Nadu, etc.

The Lateritic soils are formed under almost comparable climatic conditions as described in the preceding paragraph, but do not require alternate wet and dry conditions and the ground water level may not be very near the surface. Such soils are widely distributed in the states of Maharashtra, Andhra, Karnataka, Tamil Nadu and North-East regions, and occupy about 25 Mha of the total geographical area.

The major limitations posed by these soils are deficiency of P, K, Ca, Zn, B etc. and high acidity and toxicity of aluminium and manganese. Liming of these soils is not practical because of non-availability of lime at the sites of use and high cost of transportation. The experience of liming of temperate soils cannot be applicable in tropical areas having low activity clays. Evidences suggest that calcium is more important as a nutrient than for neutralization of acidity.

The Laterites of lower topographic positions are used for growing rice, banana, coconut and arecanut, and of higher topographic positions, for cocoa, cashew, tea, coffee, rubber etc. Shifting agriculture is mainly practiced in these areas, but the shifting cycle should be of 20 years or more.

Desert (Arid) Soils

The name Desert/Arid is given to the soils supporting negligible vegetation, except xerophytic plants, unless irrigated. Such areas may be observed in cold or hot temperature regime. A large tract of hot arid region, with a growing period of < 60 days in a year is situated in the north-west India (Rajasthan, Gujarat, Haryana and Punjab). It covers an area of about 29 Mha, and poses desertic conditions of geologically recent origin. It differs from other deserts that are cold (in the extreme north, Leh and Ladakh) and tropical (in the south in Karnataka and Andhra Pradesh) . The soils of hot arid belt are comparable with those of Alluvial soils, but have an aridic moisture regime. The Aeolian action moves and carries the sandy material and deposits it in the direction of wind in the form of a thick mantle of sand at the surface. The sandy material, under arid climatic conditions, results in poor profile development.

The major constraint of these soils is of water deficiency which restricts their use for raising agricultural crops, however the Psamments (sandy) member, situated in the interdunal valleys receiving additional rain water as runoff from the surrounding areas, are used for growing a crop (millet or pulse) during monsoon period. The Orthents are used for raising a crop on conserved moisture of monsoon rains. The Gypsids may best be used as pasture land to avoid dissolution of gypsum to form sink holes. If irrigated, all members (except gypsiferous) can be profitably used for growing two crops in a year.

Forest and Hill Soils

This name is implied for soils developed under any forest cover. In India, the total area under different forest species (tropical, deciduous, coniferous, tropical evergreen) is estimated to be 75 Mha and is observed dominantly in the state of Himachal Pradesh, Jammu and Kashimir, Uttar Pradesh, Uttaranchal, Bihar,

Madhya Pradesh, Maharashtra, Kerala and north-east region. While the climatic conditions and altitude control the kind of forest species, the kind of forest and topography control the kind of soils and their degree of profile development. The major soils observed in different forest areas are: Brown Forest and Podsolic (in northern Himalayas) and Red and Lateritic (in the Deccan Plateau). The Himalayan soils have developed on sand, limestone, conglomerates, granite, gneisses and schists under cool/cold (sub) humid climate (acidic environment), the Deccan Plateau soils are formed on igneous and metamorphic rocks (basalt, granite, gneisses) under (sub) tropical climatic conditions (slightly acidic, neutral or basic environments) and hence differ in their properties.

Podsolic Soils

The soils found under coniferous vegetation in the presence of acid humus and low base status, show some characteristics associated with Podsols, but because of the unfavourable (non-siliceous) parent material and absence of breaking down of soil minerals in the unsaturated organic acids, the process of podsolization is restricted up to the mobilization of sesquioxides and hence the true podsols in northern Himalayas (as reported by some earlier workers) are not formed.

Brown Forest Soils

The other soils, developed on sedimentary rocks and/or alluvium under sub-humid to humid climate and mixed vegetation are Non-Calcic Brown or Brown Forest soils.

The soils have great potential for growing agricultural crops such as rice, maize and fruit plants, such as apple, almond, pear, apricot etc.

Salt Affected Soils

The soils, occurring in the arid and semi-arid regions, are Intrazonal as they are interspersed with the other Zonal dominant soils of the region. According to the estimates of the Central Soil Salinity Research Institute (CSSRI), Karnal, such soils occupy 10 Mha, of which a major fraction (say 7 Mha) is sodic and occurs in the Indo-Gangetic Plain followed by the Deccan (Peninsula) Plateau supporting Black soils; the rest (about 20%), in the arid and coastal regions, are saline.

The saline-sodic soils (recently termed by the CSSRI as sodic) of the Indo-Gangetic Plain, occupy relatively lower topographic positions where products of weathering accumulate during the monsoon rains by surface runoff; evaporates and the soil solution becomes concentrated resulting in increased sodium adsorption ratio (SAR) and hence an increased ESP and pH. The displaced calcium is precipitated at high pH and temperature, as calcium carbonate. The process, repeated over years, results in the formation of sodic soils. The saline soils of the coastal region result from the rise of brackish underground water due to capillary action under excessive evaporation; the water evaporates leaving the salts at or near the surface depending upon the equilibrium established between leaching and capillary action. The salts precipitate as white efforescence and may qualify for a Salic horizon.

The soils are classified as Solanchaks (saline) and Solonetz (sodic). In terms of Soil Taxonomy, they fit in the Orders of Aridisols, Inceptisols, Alfisols and Vertisols depending upon the profile development.

The sodic soil pose serious problems of high sodium on the exchange complex, poor physical conditions, especially soil structure and drainage, nutrient and water availability and micronutrient deficiency. Another problem is of receding ground water in the central sectors and the south-west sectors of Punjab, Haryana and some parts of Rajasthan. The rise in ground water causes salinization of soils and the farmers are obliged to switch over to growing rice and eucalyptus in areas which till recently were used for raising cotton and citrus plantation.

Despite many limitations, the sodic soils, once ameliorated using gypsum technology, are used successfully for growing rice followed by wheat. The amount of gypsum, required for their amelioration would depend on sodium on the exchange complex (ESP). In saline soils, having high and brackish ground water, the objective is to reduce the soluble salts by leaching with fresh water and introduction of drainage network in order to bring the salts within the safe limit for growing crops. The presence of gypsum in such soils eliminates dispersion of clay during the amelioration process.

14

PHYSICAL PROPERTIES OF SOILS

Physical properties of soils include

- texture
- structure
- density
- porosity
- consistency
- temperature
- colour
- water content

The physical properties of a soil depend on the

- amount
- size
- shape
- arrangement and
- mineral composition of its particles
- kind and amount of organic matter
- volume
- form of its pores and
- the way they are occupied by water and air at a particular time

1. SOIL TEXTURE

Definition : Soil texture refers to the relative percentage of sand, silt and clay in a soil.

Natural soils are comprised of soil particles of varying sizes.

Texture is an important soil characteristic because it will partly determine water intake rates (absorption), water storage in the soil, the ease of tillage operation aeration status etc. and combined by influence soil fertility.

For example, a coarse sandy soil is easy to cultivate or till, has sufficient aeration for good root growth and easily wetted, but it also dries rapidly and easily loses plant nutrients through leaching.

Whereas in case of high clay soil (> 35% clay) have very small particles that fit tightly together, leaving very little pore space which permits very little room for water to flow into the soil. This condition makes soil difficult to wet, drain and till.

As the soil is a mixture of various sizes of soil separates, it is therefore, necessary to establish limits of variation for the soil separates with a view to group them into different textural classes. Texture is a basic property of a soil and it can not be altered or changed.

Soil Textural Classes

Textural names are given to soils based upon the relative proportion of each of the three soil separates – sand, silt and clay.

Soils that are preponderantly clay, are called clay (Textural class), those with high silt content are silt (textural class) those with sand percentage are sand (Textural class).

Three broad and fundamental groups of soil textural classes are recognized.

1. Sands

2. Loams (silt)

3. Clays

1. Sands : The sand group includes all soils of which the sand separates make up 70% or more of the material by weight.

Two specific classes recognized (a) sand (b) loamy sand.

2. Loams : Loamy soils containing many sub divisions does not exhibit the dominant physical properties of any of these 3 soil separates sand, silt and clay.

An ideal loam soil may be defined as a mixture of sand, silt and clay particles which exhibits light and heavy properties in about equal proportions. Note that loam does not contain equal percentages of sand, silt and clay. It does, however, exhibit approximately equal properties of sand, silt and clay.

3. Clay : A clay soil must carry at least 35% of the clay separates and in most cases not less than 40%

For e.g. Sandy clay soils contain more sand than clay. Similarly, silty clay soils contain more silt than that of the clay. Based on these 3 broad and fundamental groups, the different textural class names developed by U.S. Department of Agriculture and U.S. Bureau of soils are presented in Table1.

Table 1 : Textural class names developed by U.S.D.A.

Common name	Texture	Basic soil textural class name
Sandy soils	Coarse	Sandy, Loamy sands
	Moderately coarse	Sandy loam Fine, sandy loam
	Medium	Very fine sandy loam loam silt loam silt
Loamy soils	Moderately fine	Clay loam Sandy clay loam Silty clay loam
Clayey soils	Fine	Sandy clay Silt clay Clay

Table 2 : Textural groups on the basis of percentages of sand, silt and clay separates

Sl. No.	Textural group	Sand %	Silt %	Clay %
1.	Sand	80-100	0-20	0-20
2.	Sandy loam	50-80	0-50	0-20
3.	Loam	30-50	30-50	0-20
4.	Silt loam	0-50	50-100	0-20
5.	Sandy clay loam	50-80	0-30	20-30
6.	Silty clay loam	0-30	50-80	20-30
7.	Clay loam	20-50	20-50	20-30
8.	Sandy clay	50-70	0-20	30-50
9.	Silty clay	0-20	50-70	30-50
10.	Clay	0-50	0-50	30-100

Determination of textural class

There are generally 2 methods employed for the determination of textural class. (1) By Feel method (2) by Laboratory method

1) Feel method : In the field, texture is commonly determined by the sense of feel. The soil is rubbed between thumb and fingers under wet conditions. Sands feel gritty and its particles can be easily seen. The silt when dry feels like flour and talcum powder and is slightly plastic when wet. Clayey particles feel very plastic and exhibit stickiness when wet and are hard under dry conditions.

2) Laboratory method : A more accurate and fundamental method has been devised by the U.S.D.A. for the naming the soils based on mechanical analysis. From the Fig. (1) mentioned earlier the determination of textural names can be easily made. The diagram re-emphasizes that a soil is a mixture of different sizes of particles. This figure also suggests a gradual change of properties with change in particle size since it has not sharp line demarcation in the distribution of sand, silt and clay fractions.

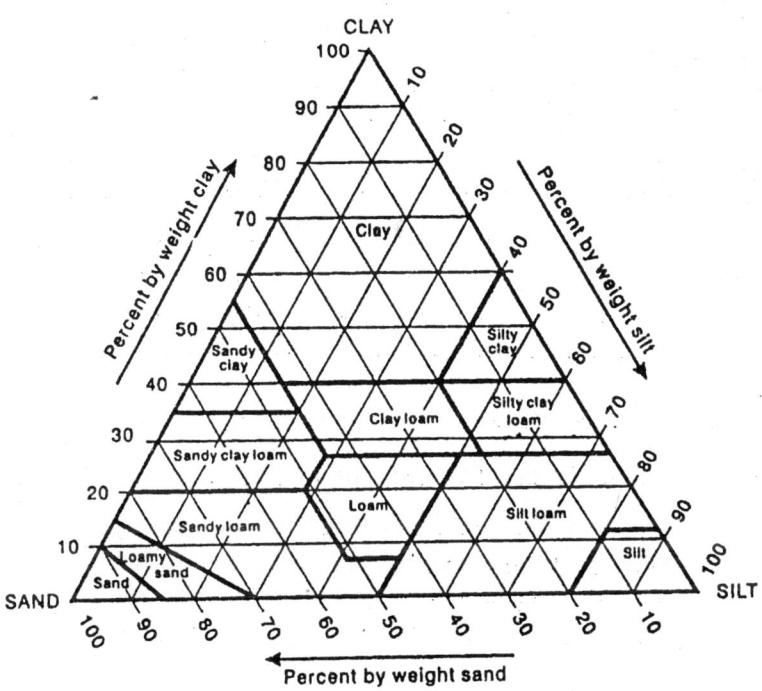

Fig. 1. Triangular textural diagram (sand, silt and clay particle sizes of 2-0.02 mm, 0.02-0.002 mm and < 0.002 mm, respectively

Classification of soil particles

A number of different classifications have been devised. They are (i) System developed by USDA (ii) system developed by British Standard Institution, (iii) system developed by International Society of Soil Science (ISSS) and (iv) system developed by United States Public Roads, Administration.

	0.002	0.006	0.02	0.06	0.2	0.6	2.0 mm
British Standards Institution	CLAY	Fine	Medium	Coarse	Fine	Medium	Coarse	GRAVEL
		SILT			SAND			
International Society of Soil Science		SILT		SAND			GRAVEL
				FINE		COARSE	

	0.002		0.02		0.2		2.0 mm

	0.002		0.05	0.10	0.25	0.5	1.0	2.0 mm
United States Dept of Agriculture	CLAY	SILT		Very fine	Fine	Med	Coarse	Very coarse	GRAVEL
				SAND				
International Society of Soil Science	CLAY	SILT		SAND			GRAVEL
				FINE ·		COARSE	

	0.005		0.05		0.25		2.0 mm

Fig. Classification of soil particles according to size by four systems

Laboratory methods : Determination of Texture.

There are several methods of mechanical analysis, of which the pipette method and Bouyocos Hydrometr Method are important. Both methods are based upon the differential rate of settling of soil particles in water, and the accuracy of the methods depends upon various conditions and assumptions.

Principles : Fundamental objective of a particle size analysis is to determine the percentage distribution of those above mentioned soil particles, sand, silt and clays in the soil mass.

The rate of fall of particles in a viscous medium depends upon the size, density and shape of the particle. In a medium like water, larger particles settle more rapidly as compared to smaller ones with the same density and consequently settle out of suspension very quickly. This principle serves as the basis of practically all mechanical analysis.

Stoke's Law : G.G. Stokes (1851) suggested the relation between the radius of a particle and its rate of fall in a liquid. He stated that the resistance offered by the liquid to the fall of the particle varied with the radius of the sphere and not with the surface.

Viscosity : It is the property which opposes the relative motion of adjacent portion of the liquid and can be consequently regarded as a type of internal friction

According to formula $\quad V = \dfrac{2}{9} \dfrac{(dp - dl)}{n} gr^2$

where,

r = radius of the particle (cm)

n = absolute viscoity of the liquid

v = velocity of fall cm/sec

g = acceleration due to gravity g/cc

dp = density of particle (g/cc)

dl = density of liquid(g/cc)

Stoke's Law

The velocity of fall of a particle with the same density is directly proportional to the square of the radius and inversely proportional to the viscosity of the medium.

This form of Stoke's law is applicable to a solid sphere or soil particle falling through a liquid or gas or to a drop of liquid falling through a gaseous medium.

Assumption of Stoke's law

1. The particles must be large in comparison to liquid molecules. So that Brownian movement will not effect the fall.

2. The extent of the liquid must be great in comparison with the size of the particle.

3. Particles must be rigid and smooth means "equivalent or effective radius" of the soil particle.

4. There must be no slipping between the particle and the liquid.

5. The velocity of fall must not exceed a certain critical value

so that the viscosity of the liquid remains the only resistance to the fall of the particle.

6. Particle greater than silt size fractions of a soil mass cannot be separated accurately with the help of the Stoke's-law.

Limitations of Stoke's law

1. The effect of different particle shapes (i.e. irregular flake, roundish, subrounded rod and disc etc.) on the settling velocities of clay particles is a major limitation for the accuracy of this law.

2. During mechanical analysis based on this principle, it is necessary to maintain a known constant temperature because the rate of fall varies inversely with the viscosity of the medium which changes with the change in temperature.

3. The density of the soil particle is another factor that affects accuracy of this law. Density of particle depends upon the mineralogical and chemical constitution of the particles and the degree of hydration. Generally, the value of the particle density 2.65 g/cc will give more or less exact figure for the determination of mechanical analysis.

Practical Implication of Mechanical Analysis

1. The mechanical analysis are not of much significance unless stone and gravel are present in large quantities exceeding 10%. If they are present beyond 10% but not exceeding too large then facilitates drainage and tillage.

2. It helps in deciding the textural class names like sand, sandy loam, clay loam etc. by determining the percentage of different size groups of particles.

3. By mechanical analysis one can easily understand the physical properties as well as colloidal behaviour of soils.

4. It can help cultivation of soil by giving an idea of light and heavy properties of soil.

The use of the terms light and heavy refers to ease of tillage and not to soil weight.

Out of 4 systems the international system for the classification of soil particles is commonly followed in India.

Soil Classes

The overall textural designation of a soil, called the textural class is conventionally determined on the basis of the mass of 3 fractions. Soils with different proportions of sand, silt and clay are assigned to different classes as shown in the triangular diagram.

e.g : soil is composed to 50% sand, 20% silt, 30% clay.

Problem

A sample of soil was sieved and had the size separates in material smaller than 2 mm determined by particle-size (Mechanical analysis) with the following results.

- Coarse fragments screened out   50 gm
  (2-50 mm diameter)
- Sand content (2 - 0.5 mm diameter)   140 gm
- silt content (0.05 – 0.002 mm diameter)   38 gm
- clay content (less than 0.002 mm diameter)   22 gm
                                                        _____
Total dry soil wt.   250 gm
                                                        _____

Determine the textural class name

Whole soil = Coarse fragments + fine earth (< 2mm)

Solution : Textural names consider only the less than 2 mm portion. The coarse fraction name is given if over 20% of the mineral soil weight is coarse material.

The coarse fraction = $\frac{50 \times 100}{250}$ = 20% and the size is gravel. Thus, the term

gravel will be added to the textural class name as determined below. Percentage of sand, silt and clay are based on only the less than 2mm fraction to substract the coarse material weight. The percentage are :

$$\frac{140}{200} \times 100 = 70\% \text{ sand}, \quad \frac{38}{200} \times 100 = 19\% \text{ silt}$$

$$\frac{22}{200} \times 100 = 11\% \text{ clay}$$

Using the textural triangle. Place the triangle with 100% clay at top and read across parallel with the base along the 11% line

11% clay }

19% silt } lines intersect in the sandy loam

70% sand (gravelly sandy loam) correct total name textural

Characteristics of soil separates/particles

A) Physical Nature of Soil Separates

Soil separates consists of stones, gravels, sands, silts, and clays

i)   Stone, Gravel and Sands (Coarse fragments). Stones and gravels range in size from 2mm upward and may be almost rounded, irregularly angular or even flat. The distinction between stone and gravel is based on size and stone (> 3 inches) is generally larger than gravel (upto 3 inches).

Sand (2 - 0.05 mm) average particle diameter may be rounded or quite irregular depending on the amount of particles attached. When sands are not coated with clay or silt, particles do not show any sticky, plastic property or any colloidal property.

Due to presence of sand particles in a soil mass, the passage of percolating water is rapid and thereby facilitate drainage and aeration. The water holding capacity (WHC) of a soil is low due to presence of good amount of sand separates.

ii)   Clay and silt (Finer fractions) : The physical properties of

128

soil are primarily dependent on total surface area. In addition to surface area, the chemical properties of the soil particles and organic matter content also exert influence.

Surface area in square centimeters per gram

or

per cubic centimeter of soil is called "Specific surface" of the soil.

Clay consists chiefly of secondary products of chemical weathering, have ultramicroscopic size, possessing large surface area than that of other fractions like silt and sand.

Clay particles generally are mica-like in shape and highly plastic when moist. When clay is wetted it tends to be sticky and is easily molded. On drying it absorbs considerable heat energy and on wetting again it evolves the same amount of heat. This phenomenon is known as "heat of wetting" clay particles exhibit properties of swelling, plasticity, cohesion and adhesion etc.

Silts are intermediate in size and show properties somewhat intermediary between sands and clays and are composed of original mineral fragments.

An unsatisfactory physical property will develop in presence of silt in higher amounts in soils and this poor physical condition can be slightly improved by supplementing adequate amount of sand, clay and org. matter to the soil.

B) Mineralogical Nature

i) Sands and silts : These are the coarse particles among other soil separates. They are fragments of rocks as well as minerals. Quartz (SiO_2) commonly dominates the finer grades of sand as well as silt separate. Besides this, there are presence of various other minerals like feldspars, mica, gibbsite, haematite, and limonite.

ii) Clays : Coarse clay particles are composed of minerals like quartz and the hydrous oxides of iron and aluminium and other aluminosilicate minerals. Three main mineral types – kaolinite, illite, and monto morillonite are mostly present.

C) Chemical Nature

i)   Sands and silt : Since quartz (SiO_2) is dominant of these two fractions. They are chemically inactive, some times these fractions especially sands, contain different insoluble nutrient elements and hence cannot supply nutrients to plants.

Silt (K-bearing minerals, micas) have been known to release potassium in soils and supply K to the plants.

ii) Clays : This fraction in soils is very active. Montomorillonite and Kaolinite are aluminium silicates. They carry Na, Fe, Mg. Illite is a hydrous mica, a potassium aluminium silicates. It contains high potassium.

SOIL STRUCTURE

Definition : The arrangement of primary particles and their aggregates into a certain definite pattern is called soil structure.

The primary particles (sand, silt and clay) do not exist as such but are bound together with varying degrees of tenacity into larger units or aggregates usually termed as secondary particles.

These are naturally occurring semi-permanent clusters or groups of soil particles, the binding forces between which are much stronger than the forces between adjacent aggregates.

Before going to discussion on soil structure we should be well acquainted with the following terms:

1.   Peds : Natural aggregates ae called peds and vary in their water stability.

2.   Clod : It is used for a coherent mass of soil broken into any shape by artificial means such as by tillage.

3.   Fragment : It is a broken ped.

4.   Concretion : It is a coherent mass formed within the soil by the precipitation of certain chemicals dissolved in percolating waters. Concretions are usually small like shot gun lead pellets.

Classification

Soil structure is described under 3 categories

I.   Types

II. Classes

III. Grades

I. Types of Structure

It is determined by the general shapes and arrangements of peds. These are mainly 4 types of soil structure (Fig.)

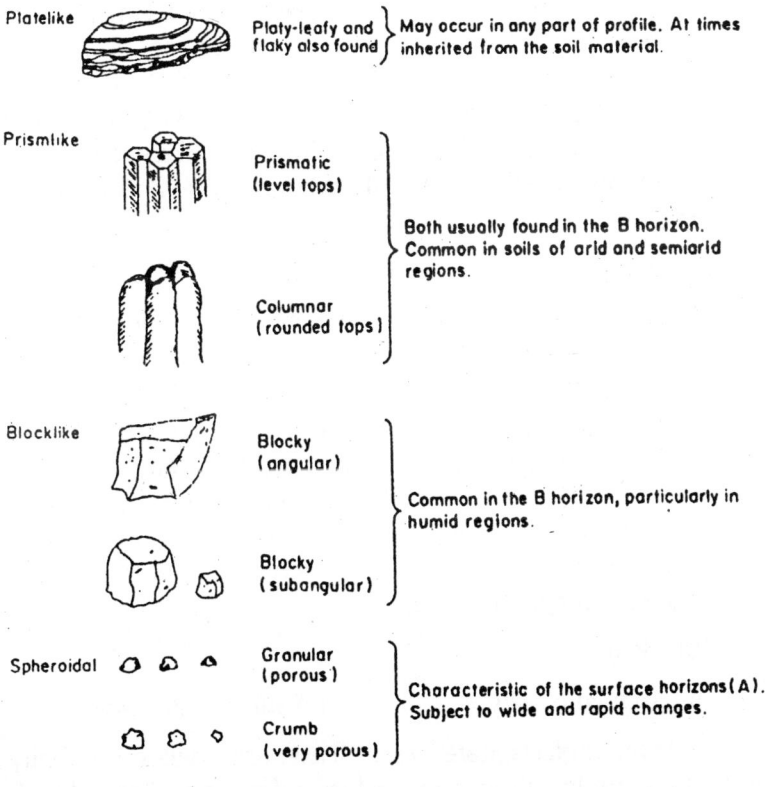

Platelike	Platy-leafy and flaky also found } May occur in any part of profile. At times inherited from the soil material.
Prismlike	Prismatic (level tops)
	Columnar (rounded tops) } Both usually found in the B horizon. Common in soils of arid and semiarid regions.
Blocklike	Blocky (angular)
	Blocky (subangular) } Common in the B horizon, particularly in humid regions.
Spheroidal	Granular (porous)
	Crumb (very porous) } Characteristic of the surface horizons (A). Subject to wide and rapid changes.

Fig. Various structural types found in soils
(*Source* : Brady and Weil, 1996)

131

(i) Plate-like (ii) Prism-like (iii) Block-like (iv) Spheroidal

(i)   Plate-like : The horizontal dimensions are much more developed than the vertical axis resulting a flattened, compressed or lens-like appearance to the peds. When the units are thick, they are called platey and when the units are thin, they are called laminar. The platey type is often inherited from the parent materials. In addition frost heaving fluctuating water tables, compaction and thin layering of different textured alluvium or lacustrine material can form platy-type of soil structure.

(ii) Prism-like : The vertical axis is more developed than others, with flattened sides, giving a pillar-like shape. It has also two sub types (1)Columnar – when the top of such ped is rounded, and (2) Prismatic when the tops of the prisms are still plane, level and clean cut. The prism like structure are commonly found in sub soil horizons in arid and semi arid regions.

Prismatic   Columnar

(iii) Block-like : All 3 dimensions are about the same size and the peds are cube-like with flat or rounded faces. Block-like structure has also 2 sub types (1) angular block - when the faces are flat and edges of the cubes are sharp angular and (2) sub - angular blocks – when the faces and edges are mainly rounded. The block-like soil structures are usually found in the sub-surface horizons and their other characteristics have much to do with soil drainage, aeration and root penetration.

Block like

(a) Angular blocky   (b) Sub-angular blocky

(iv) Spheroidal (sphere-like) : All axes are developed equally with the same length, curved and irregular faces. Generally all rounded or sphere-like peds (aggregates) may be placed in this type of soil structure. Spheroidal type of soil structure has 2 structural sub-types (1) granular-simply the aggregates of this type are usually termed as granular and it is less porous and crumly when the granules are especially porous.

The term spheroidal more appropriately refers to those sizes of aggregates not exceeding ½ inch in diameter. Granular and crumb structures are characteristic of many surface soil. In spheroid at soil structure, the physical properties of soil like infiltration, percolation and aeration etc. are not affected by wetting of soil.

    (i) Granular   (ii) Crumby

II. Classes of Soil Structure

Each primary structural type of soil is differentiated into five size classes based on the size of the individual peds.

They are as follows

1. Very fine or very thin (in case of platey soil structure)

2. Fine or thin

3. Medium

4. Coarse or thick (in case of platey soil structure)

5. Very coarse or very thick

III. Grades of Soil Structure

Grade indicates the degree of distinctness and durability of the individual peds

1. Structureless : There are no visible peds or aggregates. If the appearance is coherent as in compact clay, the term massive is used and if non-coherent as in loose sand it is called single grain.

2. Weak : Poorly formed, non durable, indistinct peds that break into a mixture of a few entire and many broken peds and much unaggregated material.

Classification of soil structure according to Soil Survey Staff (1951)

A. Type : Shape and arrangement of peds

	Plate like: Horizontal axes longer than vertical	Prismlike: Horizontal axes shorter than vertical. Arranged around vertical live vertices angular		Block like – polyhedral – spheroidal → Three approximately equal dimension arranged around a point			
				Blocklike polyhedral plane or curved surface accommodated to faces of surrounding peds		Spheroidal polyhedral plane or curved surfaces not accommodates to faces of surrounding peds	
	→ Arranged around a horizontal plane	→ Without a rounded caps	→ With rounded caps	→ Faces flattened vertices sharply angular	→ Mixed rounded, flattened faces; many rounded vertices	→ Relatively porous peds	→ non Porous peds
	Platey	Prismatic	Columnar	Blocky	blocky (Subangular)	Granular	Crumb
B Class size of peds							
1. Very fine or very thin	> 1mm	>10 mm	> 10 mm	<5mm	<5mm	<1mm	1mm
2. Fine or thin	1-2mm	10-20 mm	10-20 mm	5-10 mm	5-10 mm	1-2 mm	1-2 mm
3. Medium	2-5 mm	20-50 mm	10-20 mm	10-20 mm	10-20 mm	2-5 mm	2-5 mm
4. Course or thick	5-10 mm	50-100 mm	50-100 mm	20-50 mm	20-50 mm	5-10 mm	-
5. V. course or V. thick	>10 mm	>100 mm	>100 mm	>50 mm	>50 mm	>10 mm	-

C : Grade : Durability of peds - 0 Structure less – No aggregation or orderly arrangement

1. weak – poorly farmed, non-durable indistinct ped that break into a mixture of a few entire and many broken peds and much unaggregated material

2. Moderate – well formed moderately durable peds in distinct in undisturbed soil, that break into many entire and some broken peds but little unaggregated material.

3. Strong – well formed, durable, distinct peds weakly attached to each other, that break almost completely into entire peds.

3. Moderate : Moderately well-developed peds which are fairly durable and distinct.

4. Strong : Very well-formed peds which are quite durable and distinct.

Genesis of Soil Structure

The mechanism of structure formation is very complex. However, for the formation of aggregates the soil particles should coagulate or flocculate and should be held together or bound together into clusters by some binding and cementing materials. Aggregate formation in soil is largely a function of

a) silt and clay content

b) organic compounds present in the soil

c) the microbial activity

d) the concentration of irreversible soil colloids

e) the long range and van der walls forces acting between the charged clay minerals and the polymers present.

Besides these, many salts, parent material, climate and some of the soil forming processes also affect granulation.

Cation like Ca helps in aggregation

Evaluation of Soil Structure

Soil structure can be evaluated by determining the extent of aggregation. The stability of aggregates and the nature of the pore space. All these characteristics change with tillage practices and cropping systems. The amount and distribution of pore spaces are highly related with the aggregates and the susceptibility of the aggregates to water and wind erosion.

Aggregate analysis : There are generally 3 techniques that can be followed for the aggregate analysis

i) Wet and dry sieving

ii)   Elutriation

iii) Sedimentation

Direct dry sieving of soils in the field is used to evaluate the distribution of clods and aggregates.

Dry sieving of aggregates gives an important index for characterizing the susceptibility of soils to wind erosion.

In the wet sieving technique, the soil is slowly wetted by capillarity for 30 min. and is then transformed on to a nest of sieves immersed in water. The sieves are slowly raised and lowered in the water for 30 min. The weight of soil on each sieve is then determined.

Elutriation may be used for the separation of aggregates with diameters between 1 & 0.02 mm sedimentation methods have been used to determine the aggregate distribution in the finer fractions that cannot be separated by sieving. They are limited to aggregate sizes less than 1 mm. There are generally two limitations of sedimentation method : Varying density of particles and possibility of flocculation during sedimentation because of the downward motion of the larger aggregates.

Index of soil structure : A number of indices of soil structure have been suggested that are given below

i)   Percentage aggregates > 2.0 mm

ii)   Percentage aggregates > 0.25 mm

iii) Mean Weight Diameter (MWD)

MWD – the proportion by weight w; of a given size fraction of aggregates is multiplied by the mean (average) diameter $_i$ of the fraction. The sum of these products of all size fractions is called the mean weight diameter

$$\text{MWD} = \quad x^-_i \, w_i$$
$$i = 1$$

iv) Geometric mean diameter (GMD) – the weight of the aggregates in a given size fraction is multiplied by the logarithm of the mean diameter of that fraction. The sum of these products for all size fractions is divided by the total weight of sample to give the GMD.

v) Stability index (SI)

vi) Structural Co-efficient of soils (SS)

vii) Organic carbon

viii) Minimum bulk density (BD)

ix) Hydraulic conductivity

Methods of Evaluation : There are 4 methods of evaluation

i) Stability against disruption during wet sieving

ii) Stability against the impact of falling drops of water

iii) Stability against disintegration during leaching with dilute NaCl solutions

iv) Stability against flaking when pretreated with alcohol or other organic liquids

Among these methods wet sieving has been used extensively to determine the size distribution and the stability of aggregates.

Factors Affecting Soil Structure

i) Climate : Influences the degree of aggregation in arid region. Very little aggregation of primary particles (sand, silt & clay) In semi-arid region degree of aggregation is greater than arid regions.

ii) Organic matter : Major agent for the encouragement of granular type aggregates in soils. During decomposition of OM – Org. compounds and other slimy materials having sticky, cementing and binding properties bind the soil separates aggregation.

iii) Adsorbed cation by soil colloids influence on aggregation e.g. Na^+ Ca^{++} Cations adsorbed by

    ↓   ↓   clay colloids

deflocculated flocculated good structure

iv) Tillage : Intensive cultivation increased infiltration capacity and penetrability, but spoiled the soil structure. For obtaining good soil structure tillage operation should be made at optimum moisture conditions.

v) Type of vegetation : Grassland and forest soils have high stability of aggregates. Grasses and legume improve the aggregation of soils as compared with crops like corn.

vi) Plant roots : Secretary products from the roots of different plants may also act as cementry agent in binding the soil particles formation of good soil structure.

vii) Soil organisms : Intensive microbial activity takes place after the incorporation of organic materials to the soil. The different soil organisms like earthworm, moles, insects etc. Soil aggregation take place through their slimy and other secretary products.

viii) Manurial practices and crop rotation – cultivation of green manuring and grass crops will improve the soil structure

ix) Alternate wetting and drying – will tend to form clods and granules of various sizes.

Influence of Soil Structure on Soil Physical Properties

i) Aeration Status/Porosity e.g. Platey type – soil structure – soil aeration/ is less porosity crumby soil structure – more pore space.

ii) Temperature : Good soil structure like crumby provides a well aeration and improves the WHC of the soil. These characters help in maintaining the thermal regimes of the soil in comparison to other soil structure.

iii) Density BD – changes with changes in pore spaces –

different type of soil structure platey soil structure – low total pore spaces has high BD whereas crumby structure more total pore space has low BD.

iv) Consistence : Soil structure is influenced by the consistency of soil. Plate – like structure exhibit strong plasticity.

v) Colour –Platey structure generally hinders free drainage and percolation of water. So the soil colour changes with the soil structure. Bluish and greenish colour of the soil are generally found in poorly drained soils.

Importance of Soil Structure

1. Physical property in relation to plant growth

   - influence on porosity
   - spheroidal type – best structure – physical property

2. Soil tend to puddle under the condition of less stability of aggregation.

3. Soil structure – changed easily by different management practices namely, ploughing, draining, liming, fertilizing and manuring etc.

4. Application of O.M. improves soil structure.

DENSITIES OF SOIL

Density is the weight per unit volume of a substance or Density is usually defined as the mass of a unit volume of substance

$$\text{Density (D)} = \frac{\text{Mass (M)}}{\text{Volume (V)}} \text{ gm/cc or lb/cuft}$$

Density is expressed in grain per cubic centimeter or pound per cubic foot

Two (2) density measurements (1) Particle density and (2) Bulk density are common for soils.

Particle Density : The weight per unit volume of the solid

portions of soil is called particle density. It is expressed in gm/cc (CGS system) or /b/cft (FPS system)

In the metric system, particle density can be expressed in terms of megagrams per cubic meter (Mg/m^3) thus, if 1 m^3 (volume) of soil solids weights 2.6 mg, the particle density is 2.6 mg/m^3.

Particle density depends upon the accumulative densities of the individual inorganic constituents of the soil. Particle density depends on the chemical composition and crystal structure of the mineral particles. Generally in the normal soils the particle density is 2.65 grams per cubic centimeter 2.65 gm/cm^3.

The particle density is higher if large amounts of heavy minerals such as magnetite, limonite, haematite and zircon are present. With increase in organic matter of the soil, the particle density decreases.

Particle density of the individual soil for most mineral soil vary between 2.60 and 2.75 Mg/m^3. This narrow range occurs because of quartz, feldspar, micas, and silicate minerals usually densities make up the major portion of mineral soil.

Organic matter weighs much less than an equal volume of mineral solids.

Particle density of O.M. range between 1.1 – 1.4 Mg/m^3

$$\text{Solid particle density} = \frac{\text{Weight of solids}}{\text{Volume of solids (pore + solid)}}$$

2. Bulk density

Bulk density is defined as the mass (weight) per unit volume of dry soil (volume of solid and pore spaces). It is also expressed in gm/cubic (gm/cc) or lb/cft.

$$\text{Bulk density } Mg/m^3 = \frac{\text{wt. of oven dry soil}}{\text{volume of soil (solids + pores)}}$$

In the field 1 m³ of certain soil appears as

Solid & pore space 1.33 Mg

If all the solids were compressed to the bottom, the cube would look like

½ pore space ⟶
½ solids ⟶

To calculate BD of the soil

Volume = 1m³	Weight
(Solids +	1.33 Mg
Pores)	(Solids only)

To calculate solid particle density

Volume 0.5 m³
(Solids)
only

Wt 1.33 Mg (solid only)

$$BD = \frac{\text{wt. of oven dry soil}}{\text{Volume of the soil}} \quad \frac{(1.33)}{1}$$

therefore $BD = \frac{1.33}{1} = 1.33\ Mg/m^3$

Solid particles density $= \dfrac{\text{Wt. of solid}}{\text{Vol. of solids}}$

$$\underset{(Dp)}{Pd} = \frac{1.33}{0.5} = 2.66\ Mg/m^3$$

Bulk density of a soil is always smaller than its particle density. It is because of loose and porous soils have low weights per unit volume and compact soils have high values. The BD of sand dominated soil is about 1.7 gm/cc, whereas in organic peat soils the value of BD is about 0.5 gm cc. BD normally decreases as mineral soils become finer in texture. BD is of greater importance than particle density in understanding the physical behaviour of soils. Generally soils having low and high BD exhibit favourable and poor physical conditions respectively.

Generally in normal soils bulk density ranges from 1 to 1.60 gm/cc. In very compact subsoils = 2 gm/cc.

Factors affecting BD

(1) Amount of pore space : If the soil containing more pore spaces than that of solid spaces per unit volume than the

value of BD will be very low.

2) Compactness of the soil : In high compacted soil or waterlogged soils the BD will be more.

3) Texture of the soil : Textural variations influence the value of BD in soils e.g. clay, clayloam and silt loam surface soils show low BD as compared to sands, and sandy loam soils which show high BD value.

4) Organic matter content : Soils containing high organic matter show lower value of BD.

5) Soil structure : Soil structure affects BD by influencing the porosity of the soil e.g. crumb soil structure shows low BD than that of platey soil structure.

Problem

1. A metal cylinder pushed into a loam soil is removed from the field and the soil contains is dried in an oven. The measured data are given below.

Inside diameter of the cylinder 4.4 cm   Metal Cylinder height = 5.0 cm

Oven dry soil weight   87.6 g

Calculate the BD of the soil - ?

Solution (1) The volume of the soil sample equals the volume of the cylinder.

A cylinder's value (V) = Π r² h cubic unit

$$V = \Pi \frac{(diameter)^2}{2} \times h \text{ cubic unit}$$

$$V = 3.14 \frac{(4.4)^2}{2} \times 5 \text{ cc}$$

$$= 76.0 \text{ cc}$$

2. The bulk density of a soil can be calculated on the basis of that soil divided by its volume

$$\text{Thus, BD} = \frac{\text{Wt. of soil mass}}{\text{Soil volume}} = \frac{87.6 \text{ g}}{76.00 \text{ cc}} = 1.15 \text{ g/cc}$$

Porosity of soil

Porosity and Permeability

Porosity of soil = The volume percentage of the total soil bulk not occupied by solid particles.

Permeability of soil : The ease with which gases, liquids or plant roots penetrate or pass through a bulk mass of soil or layer of soil.

Pore - spaces (also called voids) in a soil consists of that portion of the soil volume not occupied by solids, either mineral or organic. The pore space under field conditions, are occupied at all times by air and water. Pore spaces directly control the amount of water and air in the soil and indirectly influence the plant growth and crop production. In general there are broadly 2 types of pores in soils (1) Macro pores and (2) Micro or capillary pores.

i)   Macro pores : Large sized pores are referred to as macro-pores which allow air and water movement easily. Sands and sandy soils have a large number of macro-pores. It is found in between the granules.

ii) Micro or Capillary pores : Smaller sized pores are generally referred to as a micro or capillary pores in which movement of air and water is restricted to some extent.

Clays and clayey soils have a greater number of micro or capillary pores. It has got more importance in the plant growth relationship. It is found within the granules.

Soil pores have been divided into following 4 categories based on the size grouping of soil separates.

Coarse pores : Greater than 0.2 mm or 200 microne (0.008 inch) – average diameter 1 micron = 1 millionth of a meter.

Size of Medium sands

Medium pores : 0.2 – 0.02 mm or 200-220 microns (0.008 – 0.0008 inch) average diameter size of coarse silt particles.

Fine pores : 0.02 – 0.002 mm, which 200-202 microns (0.0008 inch) average diameter size of fine silt particles.

Very fine pores : Less than 2microns (0.00008 inch) average diameter size of large clay particles.

Porosity refers to the percentage of soil volume occupied by pore spaces. Size of individual pores, is more significant than total pore space in soil in its plant growth relationship. For optimum growth of the plant, the existence of approximately equal amount of macro and micro-pores which influence aeration, permeability, drainage and water retention favourably.

Factors affecting porosity of soil

1. Soil structure : A soil having granular and crumb structure contains more pore spaces than that of prismatic and platey soil structure. So well aggregated soil structure has greater pore space as compared to structureless or single grain.

2. Soil texture : In sandy soils the total pore space is small whereas in fine textured clay and clayey loam soils total pore space is high and there is a possibility of more granulation in clay soils.

3. Arrangement of soil particles : When the sphere like particles are arranged in columnar form (one after the other) it gives the most open packing system resulting very low amount of pore spaces. When such particles are arranged in the pyramidal form it gives the most close packing system resulting high amount of pore space.

4. Organic matter : Soil containing high O.M. possesses high porosity because of well aggregate formation.

5. Macro organisms : Micro organisms like earthworms, rodents, insects etc. increase macro pores in the soil.

6. Depth of soil : With the increase in depth of soil, the porosity will decrease because of compactness in the sub soil.

7. Cropping : Intensive crop cultivation tends to lower the porosity of soil as compared to fallow soils. The decrease in porosity may be due to reduction in organic matter content.

8. Puddling : Due to puddling under sufficient soil moisture, the soil surface layer is made dense and compact. Due to compactness porosity of soil surface is reduced by the infiltration of muddy surface materials.

Relationship between Porosity and Densities of soil

% solid space $= \dfrac{BD}{PD} \times 100$

Since, % pore space + % solid space = 100

or % of pore space = 100 - % solid space

or % pore space $= 100 - \dfrac{BD}{PD} \times 100$

% pore space $= 100 \left(1 - \dfrac{BD}{PD}\right)$

For example, a soil having bulk density of 1.5 and particle density of 2.65 calculate the porosity of soil

% pore space $= 100 \left(1 - \dfrac{1.5}{2.65}\right)$

Porosity of the soil = 43.40%

Specific surface : Area in cm^2/g or cm^3/g of soil is called specific surface of soil.

This is one of the common property of inorganic and organic colloid is their extremely smallsize colloids are too small to be seen with an ordinary light microscope.

Most of the colloids are smaller than 0.002 mm in size (diameter).

Because of their small size, all soil colloids expose a large external surface per unit mass.

The external surface area of 1 g of colloidal clay is at least 1000 times that of 1 g of coarse sand. The total surface area of soil colloids ranges from 10 m^2/g for clays with only external surfaces to more than 800 m^2/g for clays with extensive internal surfaces. The colloidal surface area in the upper 15 cm of a hectare of a clay soil could be as high as 7,00,000 km^2.

Specific surface area of some clay minerals

Clay minerals	Range of surface area (m^2/g)		
	Internal	External	Total
Kaolinite	7-10	30-35	37-45
Illite	70-100	50-70	100-170
Montmorillonite	500-600	80-150	580-750
Chlorite	60-80	70-100	130-180
Vermiculite	700-800	80-100	780-900

Plasticity and Cohesion

Soil colloidal particles may present in gel condition possess the property of plasticity. Due to this property clay colloids are moulded in any shape.

Soils containing > 15% clay exhibit plasticity. This property is probably due to the plate-like nature of the clay particles and the combined lubricating and binding influence of the adsorbed water. The particles of the plastic soils easily slide over each other, much like panes of glass with films of water between them.

Plasticity exhibited over a range moist to wet soil levels refered to as plasticity limits. The lower of these levels termed plastic limit of the soil. The plastic limit is the lowest moisture content at which a soil cannot be deformed without cracking soils. Soils should not be tilled at moisture contents above the plastic limit.

The upper plastic limit or liquid limit is the moisture content at which soil ceases to be plastic, becomes semifluid (like softened butter) and tends to flow much like liquid. While these limits have only modest use for agricultural purpose.

Plasticity is of practical importance because of its influence on tillage operations. Thus, the cultivation of a fine textured soil when it is too wet will result in puddled condition detrimental to suitable aeration and drainage with clayey soils, especially those of the smectite type, plasticity presents a significant problem. Stable granular structure is often difficult to establish and maintain in clayey soils high in smectites.

Cohesion :

Unlike sand, clay particles possess the properties of cohesion and adhesion. While farming aggregates, the colloidal clay particles unite each other by the virtue of the property of cohesion. Clay particles envelop sand particles under the force of adhesion are developed in the presence of water. When colloidal substances are wetted water first adheres to the particles and then brings about cohesion between two or more adjacent colloidal particles soil. When dried, the particles remain united because of the force of molecular cohesion. These two forces helps in the retention of water in the soil and thus used by plants and microorganisms.

Swelling and Shrinkage :

Soil particle when brought in contact with water they imbibe a certain quantity of water and swell and increase in volume.

Some clays such as smectite or montmorillonite swell when wet and shrink when dry. After prolonged dry spell, soils high in smectite (Vertisols) often are criss crossed by wide deep cracks.

These properties of soils are largely responsible for the development and stability of soil structure.

Swelling and shrinkage are more dominant in montmorillonite clays (2:1 expanding type mineral) than that of Kaolinite (1.1 non expanding type mineral) clays.

Flocculation and deflocculation

Flocculation : The colloidal particles are coagulated by adding an oppositely charged ion. Formation of floc is known flocculation. If the cations are held close to the negatively charged particles, the negative charge would be neutralized and colloidal particles flocculate and settle down Na^+ (sodium cation) are highly hydrated and are monovalent they are not so closely bound with the negatively charged immobile particles. Thus, the particles continue to offer resistance to aggregation and do not flocculate Ca^{++} (Calcium cations) are divalent and are not easily displaced as sodium. Thus, calcium ions are able to neutralize the negative charge more efficiently and the colloidal system tends to flocculate. In a similar manner trivalent ions like Al^{+++} are still more efficient in flocculation of colloids. Thus Na^- clay produces deflocculation and Ca^- clay encourages aggregation. The phenomenon of flocculation plays an important part in the cultivation of sils. When clay particles are flocculated soils develop small clods of a crumby nature. Such soils allow free movement of air and water. If the particles are deflocculated the aggregates get dispersed, the soil gets water logged, the movement of air and water is restricted.

Soil aggregate Stability and Soil tilth

The stability of aggregates is of great practical importance. Some aggregates readily destroy to the beating of rain and the rough plowing and tilling the land. Some of the soil aggregates resist disintegration. Three major factors appear to influence aggregate stability.

1. The temporary mechanical binding action of micro-organisms especially the thread like filaments (mycelia) of fungi. These effects are pronounced when organic matter is added to soils and are at a maximum a few weeks or month after the application.

2. The cementing action of the intermediate products of microbial synthesis and decay, such as microbially produced gums and certain polysaccharides.

3.   The cementing action of the more resistant stable humus components aided by similar action of certain inorganic compounds, such as iron oxides. These materials provide most of the long term aggregate stability.

Aggregate stability is not entirely an organic phenomena. There is continual interaction between organic and inorganic components. Polyvalent inorganic cations cause flocculation (Ca^{++}, Mg^{++}, Fe^{++}, & Al^{3+}) to provide mutual attraction between the organic matter and soil clays, encouraging the development of clay organic matter complexes. In addition films of clay called "Clay skins" often surrounded the soil peds and help provide stability.

Tilth depends not only on granule formation and stability, but also on such factors as BD, soil moisture content, degree of aeration, rate of IR, drainage etc.

Tilled the soil for 3 primary reasons (1) To control weeds, (2) to present a suitable seed bed for crop plants, (3) to incorporate organic residues into the soil. The long-term effects of tillage especially plowing, are generally undesirable. Rapid breakdown of organic residues can hasten the reduction of soil organic matter content and with break down of soil aggregates soil loose aggregate stability. Minimum the need for soil tillage.

Intensive cultivation increased infiltration capacity but spoiled the soil structure.

15

SOIL WATER

The Liquid Phase of Soil

Soil water relationships for several reasons

1. Large quantity of water must be supplied to satisfy the requirements of growing plants because water is continually lost by evaporation from leaf surfaces. Thus water must be available when the plants need it, and most of it must come from the soil.

2. Water is the solvent that together with the dissolved nutrients makes up the soil solution from which plants absorb essential elements.

3. Soil moisture helps control two other important factors essential to normal plant growth – soil air and soil temperature.

4. Water also cause for soil erosion – excess flow of water lifts soil particles from the soil surface and carries them into streams, lakes and the oceans.

A representative cultivated loam soil contains approximately 50% solid particles (sand, silt, clay and organic matter), 25% air and the rest 25% water. Only half of this water is available to plants because of the mechanics of water storage in the soil.

Structure of Water

Water is a simple compound, its individual molecules containing one oxygen atom and two much smaller hydrogen atoms. The elements are bounded together (H-O-H) covalently, each hydrogen atom sharing its single electron with the oxygen the hydrogen atoms are attached to the oxygen in sort of a 'V' shaped arrangement at an angle of only 104.5°.

Polarity : Due to 'V' shaped structure of water the side on which the hydrogen atoms are located tends to electropositive and apposite side electronegative. Because of non-linear position of H^+ ions, water is polar.

Polar means there is no center of zero change from which there is an equal change at some distance from that center in all directions (Molecules whose +ve and -ve charge center do not coincide are termed polar molecules). The H of water in soils may bound to oxygen ions of soil mineral surfaces, thereby holding the water tightly to soil. This accounts for the polarity of water and therefore, water is most important for carrying out many reactions in soils and plants (due to polarity of water). Strong combined adhesion and cohesion forces cause water films of considerable thickness to be held on the surface of soil particles. Because the forces holding water in soil are surface attractive forces, the more surface (more clay and org. matter) a soil has, the greater is the amount of adsorbed water.

So soil holds water in 2 ways in the pores or capillaries between the solid particles, and by adsorption on the solid surfaces of the clay and organic matter. The mechanism of adsorption of water on the soil surfaces are related to the adhesion and cohesion forces through hydrogen bonding and also related to the hydration of exchangeable ions which may result in some of them dissociating from the surface into the water.

The effect of the cation on the water molecules is greater the greater its change and the smaller its size, so the greater its surface charge density, and these effects are influenced by the relative moisture content of the clay, by the heat evolved during wetting of the clays and by the greater apparent density of the clays in water.

Heat of solution : When ions are hydrated, a large amount of energy is released and this is known as heat of solution.

Heat of wetting : When clay particles are hydrated a certain amount of energy must be released and this phenomenon is known as heat of wetting. So there is a close relationship between moisture retention in soil and the energy. The force with which water is held is also termed as suction. Soil water is held by adsorption and capillary forces.

Surface tension is an important property, especially as a factor in the phenomenon of capillarity.

The phenomenon of surface tension is generally evidenced at water air interfaces and it may be defined as the forces in dryness acting at right angle to any line of 1 cm length in the surface (1 cm length in the surface).

At the surface, the attaction of air for the water molecules is much less than of water molecules for each other. Consequently there is a net downward (inward) force on the surface molecules, resulting in sort of a compressed film at the surface. This phenomenon is called surface tension.

Soil Water Energy Concepts

The retention and movement of water in soils, its uptake and translocation in plants and potential evapotranspiration etc. are also related to energy. Different kinds of energy are involved including potential, kinetic and electrical. By using the term 'free energy' (ability to do work) energy status of water can be characterized. To indicate the strength with which water is held. Several concepts have been used.

The concept of pressure – the pressure required to force the water off soil and was measured in atmosphere of pressure needed to remove water. The opposite of pressure moisture suction or tension.

Recently soil water potential is used and it may be defined as the work the water can do when it moves from its pressure state to a pool of water in the reference state. The movement of water in soils takes place from a higher free energy to lower free energy level. Adsorbed water is less free to move as compared to water in a pool.

Adsorbed water always less free energy (less ability to do work) than water in the pool (zero potential). Therefore, adsorbed water always has a negative potential; work must be done to remove the water to a free pool of water. The more tightly water is adsorbed, the more negative is the number. The soil water potential is a combination of the effects of the surface area of soil particles and small pores that adsorb water.

1. Matric Potential (Ψ_m) the effects of attraction of ions and other solutes for water, (solute - in a solution the dissolved substance is called the solute)

2. Solute or Osmotic Potential (Ψ_s) and the atmospheric or gas pressure effect; pressure potential (Ψ_p). In salt, free well drained soil, matric potential is almost equal to the soil water potential (Ψ_w).

An additional effect of the position of the water (such as being elevated) compared to the reference state (the reference free energy state $= 0$ and is at a specified elevation) is called the Gravitational Potential (Ψ_g). Gravitational potential is not related to soil properties, only to the elevation of water in comparison to a reference position.

Various potentials can be written as follows :

Ψ_w	$=$	Ψ_m	$+$	Ψ_s	$+$	Ψ_p
Soil water potential		matric potential		solute or osmotic potential		pressure potential

Ψ_t	$=$	Ψ_2	$+$	Ψ_g
total water potential		Soil water potential		Soil gravitational potential

Most of the productive soils have no depth of water standing on it and can be written as follows

$$\Psi_t = \Psi_w = \Psi_m$$

Therefore, among all potentials matric potential (ψ_m) is the most important and dominant for most soils.

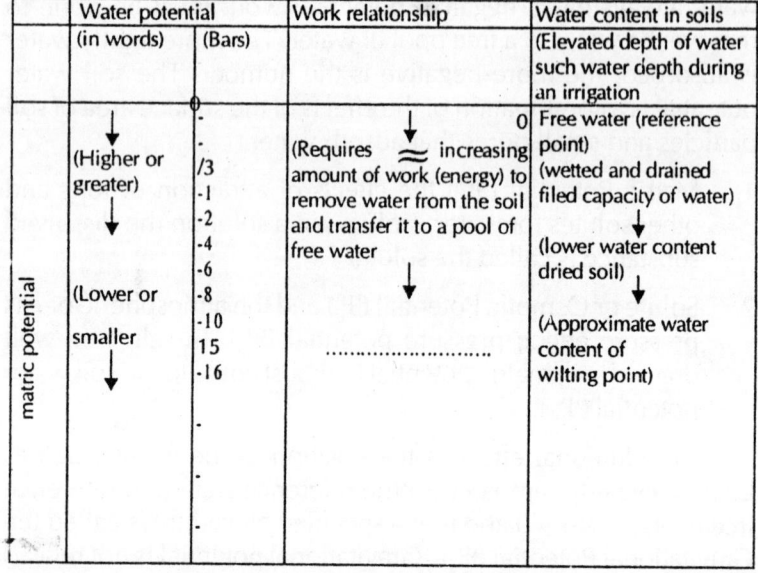

	Water potential		Work relationship	Water content in soils
	(in words)	(Bars)		(Elevated depth of water such water depth during an irrigation
matric potential	↓ (Higher or greater) ↓ (Lower or smaller) ↓	- /3 -1 -2 -4 -6 -8 -10 15 -16 - - -	↓ (Requires ≈ increasing amount of work (energy) to remove water from the soil and transfer it to a pool of free water ↓	0 Free water (reference point) (wetted and drained filed capacity of water) ↓ (lower water content dried soil) ↓ (Approximate water content of wilting point)

Fig. : A relationship between water potential, and water content in soil

Water with a high potential has more mobility in a soil than has water with lower potential. The matric potential is zero at saturation and does not ever become a positive number.

Methods of expression of soil water potential or suction

Soil water potential can be measured in two units at varying energy levels in soil.

1. PF scale : The free energy is measured in terms of the height of a column of water required to produce necessary suction or water potential at a particular soil moisture level. So the PF may be defined as the logarithm of centimeter height of water column to give the necessary suction (potential). Here 'p' indicates the logarithmic value and 'F' indicates free energy. e.g. PF 4 is equal to 10,000 centimeters of a water column height (logarithm of 10,000 = 4).

2. Bars or Atmospheres : Atmosphere or Bar is the average air pressure at sea level.

The term milli bar (m bar) = $\frac{(1)}{10{,}000}$ atmosphere

A unit bar is equated to a number of other units as follows :

1 bar = 0.9869 atmosphere (approx. 1 at m)

= weight of a 1020 cm water column

= 14.5 lbs per sq. inch

= 10^6 drynes per sq.cm

= 75.01 cm height of mercury column

If the suction is very low as occurs in case of a wet soil containing large amount of water that it can hold, the pressure difference is of the order of about 0.01 atmosphere or 1.0 pF equivalent to 10 cm height of water column. Similarly if the pressure difference is 0.1 atmosphere the PF will be 2.0.

Table : **Relationship among pF values, height of waters column and pressure/atmosphere**

Height of water column (cm)	Pressure (atmosphere or bars)		PF values
1	0.001	↑	0 saturated soil
10	0.01	Gravitational	1
10^2	0.1	water	2
346	1/3	↑	2.53 field capacity
10^3	1.0		3
10^4	10.0	↓	4
15,849	15.00	2) Capillary	4.18 wilting point
31,623	31.0	water ↓	4.50 Hygroscopic point
10^5	100.00	Hygrosopic	5
10^6	1,000.00	water	6
10^7	10,000.00	↓	7 oven dry soil

A saturated soil has PF value 0,while an oven dry soil has a PF 7.0.

Classification of Soil Water

There are generally 2 types of soil water classification based on drying of wet soils and growing plants therein.

(A) Physical classification and (B) Biological classification

A) Physical Classification

Under physical classification soil water is grouped into 3 on the basis of retention

1) Gravitational water   (2) Capillary water
(3) Hygroscopic water

1. Gravitational Water : Gravitational water may be defined as the water that is held at a potential greater than – 1/3 bar (saturated soil) and that portion of the soil water that will drain freely from the soil by the force of gravity. Inspite of having low energy of retention, gravitational water is of little use to plants water occupies the larger pores resulting poor aeration. Therefore, the removal of excess water is a must for the growth of most plants.

2. Capillary Water : Capillary water is held in the micropores of soils (Capillary pores). Capillary water is retained on the soil particle by force of attraction between soil particles and water molecules.

    Capillary water is held so rigidly that the force of gravity is not able to separate it from the soil particles. Capillary water is free and moves through the soil pores because of a water potential gradient. Capillary water may be defined as the water that is retained in the soil between the water potential of –1/3 bar to – 31 bars (Field capacity to Hygroscopic point). The force of retention of water molecules by the soil particle is high and part of water is available and part of it is unavailable and so all capillary water is not available to plants.

3. Hygroscopic Water : Hygroscopic water is defined as the water that is held by the soil particles at a suction of more than –31 bars (more than Hygroscopic point). It is essentially non liquid and moves primarily in the vapour form. This water is held so tenaciously that plants are not able to absorb it and thereby unavailable to plants. Some micro-organisms can utilize such form of water.

B) Biological Classification

There is a definite relationship between moisture retention and its utilization by plants. Biological classification is based on the availability of soil moisture to the plants. Soil water under this sytem of classification can be divided into 3 (three) categories.

1. Available water : Available water is defined as that portion of water which is retained in the soil between field capacity (–1/3 bar) and the permanent wilting coefficient (–15 bars). This water is easily available by plants and therefore, it is called plant available water. Plant available water is equal to the difference of water percentage at field capacity and permanent wilting point.

2. Unavailable water : Unavailable water is defined as the water which is held at soil water potential greater than –15 bars. It is unavailable to plants. It includes the whole of the hygroscopic water plus a part of the capillary water below the wilting point.

3. Superfluous water : Superfluous water is defined as the water which is retained in the soil beyond the field capacity soil moisture tension. This water includes gravitational water plus a portion of capillary water removed from large interstices (pores) such type of water is unavailable to plants and rather presence of such water in the soil for a long period causes harmful effect for plant growth because of lack of air.

Soil Moisture Constants

Soil moisture constants and their approximate equivalents in base of water potential as they affect the relative availability of water to plants are shown in Fig.

Oven dry	Air dry		Hygroscopic water	Permanent wilting point	Field capacity	Saturation
•		Unavailable water		Plant available water	Gravitational water drainable	Flooded condition
	-10,000	-1,000	-31	-15	- 1/3	0

⟶ Increasing water content ⟶

Fig. : Soil water constants and their approximate equivalents in bars of water potential as they affect the relative availability of water to plants.

At water potential of –1/3 bars, water is held too loosely to overcome the effect of gravity and drains away.

Capillary water (that held by capillary pressure) remains for plant use; this is held by water potentials ranging from –1/3 bar (field capacity to –31 bars or lower (Hygroscopic), depending upon the pore size of the soil. Plants can only use capillary water held by not more than –15 bars (the point of permanent wilting). Soil water at the air dry state is held by water potentials that vary from –1,000 to –10,000 bars, depending on humidity.

Soil moisture constants

1. Oven dry weight : Oven dry weight is the basis for all soil moisture calculations.

    The equilibrium tension of the moisture at oven dryness is 10,000 atmosphere or bars (-10,000 bars of soil moisture potential). It is determined by placing the soil in an oven at 105°C unit it loses no more water.

2. Air dry weight : Air dry weight is a some what variable term, mainly because the moisture in the air fluctuates. Moisture at air dryness is held with a force of 1,000 atmospheres of

bars (–1,000 bars of soil moisture potential). This water is not available to plants.

3. Hygroscopic coefficient : Hygroscopic coefficient is determined by placing an air dry soil in a nearly saturated atmosphere at 25°C until soil absorb no more water. The soil moisture tension at this point is equal to 31 bars (soil moisture potential –31 bars) and this water is not available to plants, but available to certain microorganisms.

4. Wilting co-efficient : Some times it is also used as permanent wilting point.

The wilting point is defined as that amount of water which is held with water potential less than –15 bars and it is held so strongly that plants are notable to absorb it for their needs. At this point of soil moisture potential, the plants begin to wilt and at the very beginning of the wilting condition is some times recovered the addition of water and it is then called temporary wilting point, while such wilting condition of the plant is not recovered inspite of addition of water and then it is called permanent wilting point. Both the wilting point indicate low moisture availability to plants.

5. Field capacity : Field capacity is defined as the capacity of a soil to retain moisture against the downward pull of the force of gravity and moisture is held with soil water potential less than –1/3 bar. It is used to determine the amount of irrigation water needed and the amount of reserve soil water available to plants.

Moisture equivalent is approximately equal to the amount of moisture held at field capacity soil. The term moisture equivalent is defined as the percentage of water held by a one centimeter thick moist layer of soil after subjected to a centrifugal force of 1,000 times gravity for half an hour.

Another term "Maximum water holding capacity or maximum water retention capacity" is also used. MWHC is defined as the capacity of a soil to retain water is exceeded. At this point all soil pore spaces (macro and micro pore spaces) are filled up with water and the drainage is restricted. The water at this point is at a low soil moisture tension. Under natural field conditions only poorly drained soils are at their maximum water holding capacity for long periods of time.

Factors affecting gravitational, capillary and hyroscopic water

Gravitational water : (1) Soil texture (2) Soil structure – are two main factors that effect the amount of gravitional water.

Texture : It plays an important role in regulating the flows of gravitational water. The flow of water is directly proportional to the size of the particles. The larger the size of the particle, the more rapid is the movement of water.

Structure : The different types of soil structure affects the gravitational water by influencing its movement as well as drainage condition of soils e.g. the rate of movement of gravitational water is slow in platey soil structure which results stagnation of water on the soil surface. Whereas, spheroidal soil structure helps to improve the movement of gravitational water by increasing its rate of infiltration and percolation. Hard pans in the sub soil horizon compactness of soil, O.M. contact in soil etc also affect the amount and rate of movement of gravitational water.

Capillary water : Soil texture, structure, surface tension, O.M. content, size of capillary pores in soil, for tortuosity (zig-zag path) of capillary soil pores etc.

Soil texture : Finer texture of soil – larger quantity of water it holds. Due to greater surface area and a large number of micro-pores spaces present in such soil.

Soil structure : Type of soil structure hold water varying quantities e.g. platey soil structure hold excess water than that of granular soil structure.

Surface tension : An increase in surface tension increases

the amount of capillary water. Surface tension is therefore an important property and factor that influence the movement and amount of water in the phenomenon of capillarity.

O.M. content : The presence of O.M. in the soil increases the percentage of porer space and consequently increase the capillary capacity of soil. O.M. also influences the soil aggregation as well as formation of soil structure which also affect the amount of capillary water. Humus has a greater capacity for holding water especially capillary water.

Sizes of soil pores : Different sizes of soil pores hold water with different tenacity. Small and medium sized soil pores tend to hold water with much more tenacity than that of larger size soil pores. So the movement of capillary water is largely dependent upon the size of capillary pores since different energy levels are associated with water present in different sizes of pores. Therefore, it affects the availability of such capillary water to the plants.

Tortuosity : (Zig-zag path) – Soil pores are not continuous, straight and uniform like that of capillary glass tubes. Due to such nature of soil capillary pores, the movement of water is some what restricted and different soil pores are filled with air which may be entrapped slowing down or preventing the movement of capillary water.

Hygroscopic water : The smaller the size of soil particles the greater the amount of hygroscopic water it adsorbs. Fine texture soils, like clay, clay loam soils contain more hygroscopic water as compared to coarse-textured sandy soil, clay minerals of mantmorillaite type having large surface area adsorb more water than of Kaolonite type of clay minerals

Factors affecting available water is influenced by a number of factors like plant, climatic and soil factors.

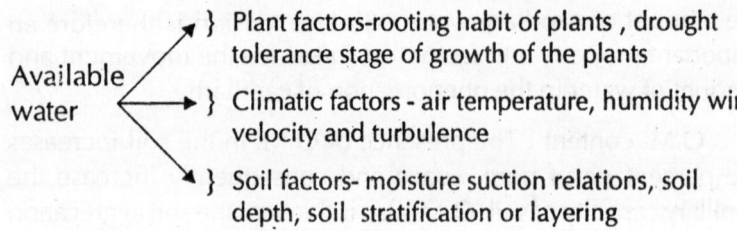

Available water
} Plant factors-rooting habit of plants , drought tolerance stage of growth of the plants

} Climatic factors - air temperature, humidity wind velocity and turbulence

} Soil factors- moisture suction relations, soil depth, soil stratification or layering

The plant and climatic factors are related to the losses of water vapour under the system known as SPAC (Soil Plant Atmosphere Continuum).

Measurement of soil moisture

1. By gravimetric method

2. Electrical conductivity method

3. Measurement by using Tensiometers (it measures metric potential in situ / field)

4. Neutron scattering method – Neutron probe – look like flash light cylindrical with long cards - probe contain radio active material (radium that emits moving neutrons)

Soil moisture release characteristics and hysteresis

Hysteresis : An interesting phenomenon occurring when the soils are alternately wetted and dried (Fig.).

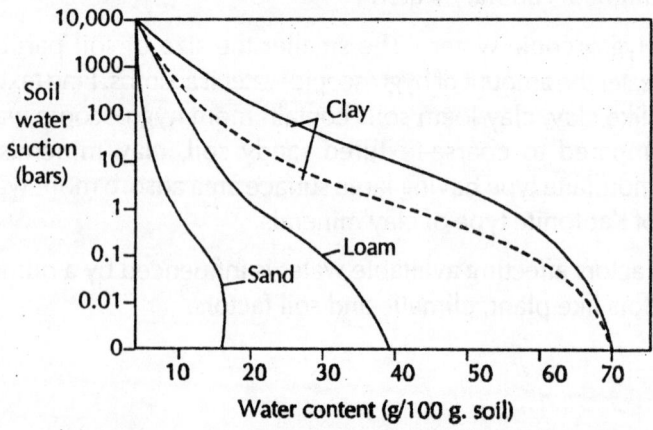

The upper solid line is termed as desorption curve develops due to drying of a saturated soil. The lower dotted line is termed the sorption curve develops due to wetting an initially dry soil. The difference between the two curves is due to the phenomenon of hysteresis resulting from the presence of entrapped air in such soil.

Hysteresis phenomenon exists in soil due to shrinking and swelling. Shrinking and swelling affect pore size on a microbasis as well as on the basis of overall bulk density. So hysteresis phenomenon is caused by a number of factors like shape, and size of soil pores and their interconnection with each other. Pore configuration, nature of soil colloids, BD of soil and entrapped air etc. Most important factor affecting hystereosis is the entrapment of air in the soil under rewetting condition. This prevents effective contact between others.

Soil Water Movement or Movement of Soil Water

There are mainly 3 types of movement namely : (1) Saturated flow (2) Unsaturated flow and (3) Water vapour movement.

1. Saturated flow : Water moves because of water potential gradients in the soil caused mostly by gravity, salt content and water usage and direction of flow is from a zone of higher to zone of lower moisture potential.

    When soil water moves mainly due to gravity, which is at moisture potential greater than $-1/3$ bar, the movement is called "Saturated flow".

    Saturated flow starts with water infiltration, which is the movement of water into the soil when precipitation of irrigation water is on the soil surface. When the soil profile is completely saturated with water, the movement of max. water flowing through the saturated soil is termed percolation.

    The flow of water under saturated conditions is determined by two major factors the hydraulic force driving the water through the soil and the ease with which the soil pores permit water movement.

V = Kf

where V = the total volume of water moved per unit time

K = hydraulic conductivity of soil or permeability

f = water moving force

Darcy stated that the rate of flow was increased with an increased depth of water above the bottom of the soil through which is flowed. The flow decreased with an increased depth of soil through which water flowed. The hydraulic conductivity (K) will be different for different types of soils.

Factors affecting saturated flow of water

1. (1) Texture (2) Structure (3) O.M. (4) inorganic colloids (2:1 or 1:1 type) (5) Pressure : Soil air pressure increased and percolation decreased and this usually found in the sub-soil horizons.

2. Unsaturated soil : It is the flow of water held with water potentials lower than −1/3 bar. Water will move toward the region of lower potential (towards the greater "pulling" force).

In a uniform soil this means that water moves from wetter to drier areas. The water movement may be in any direction. The rate of flow is greater as the water potential gradient increases and as the size of water filled pores also increases.

The 2 forces responsible for this movement are the attraction of soil solids for water (adhesion) and capillary. Under field conditions this movement occurs when the soil macropores are filled with air and micropores with water and partly with air.

Factors affecting the unsatured flow

Similar way to that of saturated flow. Amount of moisture in the soil affects the unsaturated flow. The higher the percentage of water in the moist soil, the greater is the suction gradient and more rapid is the movement (release).

3. Water Vapour Movement

The movement of water vapour form soils takes place in 2 ways (1) Internal movement – the change from the liquid to the vapour state takes place within the soil, that is in the soil pores and (2) External movement – the phenomenon occurs at the land surface and the resulting vapour is lost to the atmosphere by diffusion.

Diffusion : mechanism – takes place from one area to other soil area depending upon the vapour pressure gradient (moving force).

There are mainly 2 soil conditions that affect the water vapour movement (1) Moisture regimes (2) Thermal regimes and others are O.M. vegetative cover, soil colour etc. also affect the movement of water vapour.

In dry soil some water movement takes places in the vapour form and such vapour movement has some practical implications in supplying water to drought resistant plants.

Losses of soil water and their management the losses of soil water through percolation and evapotranspiration process.

1.   Percolation losses – maximum under humid climate where precipitation (RF) is high. Essential plant nutrients are lost from soil through such percolating water.

This loss can be reduced by application of O.M. in soil – especially moderate to high amount of sand.

2.   Evapotranspiration (E) – Two ways (loosses evaporation)

(1) at the soil surface   (2) transpiration from leaf surface

Combined loss (ET) – responsible for most of the water removal from soils under normal field conditions. The rate of ET increases when the air is dry (low relative humidity), warm, or moving winds and if the soil water is near the field capacity to be absorbed by roots or evaporated from the soil surface.

ET – losses can be reduced by selecting more growth efficient plants, mulching (sludging) or cooling the area or by using moisture barrier.

Consumptive use and water use efficiency

Consumptive use is the quality of water lost by ET plus that contained in plant tissues CUW = ET. WUE–the amount of water (transpiration, plant growth, evaporation from soil drainage loss) required to produce a unit of dry weight material (a Kilogram of corn) is a measure of WUE.

WUE = $\underline{DW}$ – Dry Weight of crop per acre

   ET   – Water used in ET in acre inches/acre

16

SOIL AIR

Total volume of a soil – One fourth of the total volume of soil is occupied by air (25% of soil air and 25% of soil water). There is interrelationship of a soil air and soil water.

Definition of Soil Aeration : defined as the exchange of carbondioxide and oxygen gases between the soil pore space and the aerial atmosphere.

So, a well aerated soil is one in which gases are available to growing aerobic organisms in adequate amounts and in proper proportions to encourage optimum rates of the essential metabolic process of the aerobic organisms.

There are 2 types of pores involved in the soil aeration – pores between the crumbs called intercrumb pores and pores within the crumbs called crumb pores.

Reasons for poor aeration

There are generally 2 reasons by which poor aeration results.

1. Excess moisture when a soil is subjected to excess moisture, waterlogged condition is developed. This situation is generally found on poorly drained, fine textured soils having a minimum of macro pores through which water can move very rapidly. In poor soil aeration – most of the plants can not grow. Poor aeration can be prevented by the removal of excess water by drainage.

2. Gaseous interchange : The inadequate interchanges of gases between the soil and the free atmosphere depends on 2 factors.

1. The rate of biochemical reactions influencing the soil gases and

2. the actual rate at which each gas is moving into or out of the soil.

The greater exchange of gases would be required because of rapid consumption of O_2 and release of CO_2 in soil.

The exchange of gases between the soil and the atmosphere is facilitated by two mechanisms (1) Mass flow (2) Diffusion.

1) Mass flow – due to pressure differences between the atmosphere and the soil air.

Temperature may influence the renewal of soil air by 2 ways (1) There may be temperature variation within the soil between horizons. The contraction and expansion of the air within the pore space as well as the tendency for warm air to move upward may cause some exchange between the different layers and with atmosphere. (2) The soil and the atmosphere have different temperature.

2) Diffusion : is the molecular transfer of gases. The molecules of gases are in a state of movement in all direction.

Composition of soil air

Name of gas	Percentage by volume	
	Soil air	Atmospheric air
Oxygen	20.00	21.00
Nitrogen	78.60	78.03
Carbondioxide	0.50	0.03
Argon	0.90	0.94

Composition of soil air important for crop growth as nutrients and water.

Aerobic respiration in roots – microorganisms and soil fauna involve the continuous consumption of O_2 and the evolution of CO_2.

Factors affecting composition of soil air

1) Physical properties, soil condition, type of vegetation, seasons, amount of O.M. and microbial activity, depth of soil and temperature etc.

Oxygen : less in soil air than atmospheric air. Plant roots, microorganisms require oxygen for their metabolism (growth) which they take O_2 from soil air and thereby decrease the concentration of O_2 in the soil environment.

Depth of soil – increase in depth the amount of O_2 is less. Due to more slow diffusion of O_2 from the sub soil horizons to the atmosphere through overlying soil layers.

Seasonal variation : affects composition of soil air. In dry season/summer season the quantity of O_2 is usually higher than that of rainy season. Because in the summer season the opportunity for the gaseous exchange is greater as compared to monsoon season resulting high O_2 and low concentration CO_2 in the former season.

Cultural and other soil management practices affect the composition of soil air by modifying physical properties of soil like BD, porosity, soil structure etc.

CO_2 : During decomposition of organic matter CO_2 is evolved concentration will be very high in O.M.

Oxygen Diffusion Rate (ODR)

It is one of the criteria generally used to determine the O_2 concentration in the soil pore space.

It consisted of allowing the free diffusion of O_2 into a diffusion chamber that was inserted into the soil. The calculated partial pressure of O_2 at 10 min. interval was used to evaluate the diffusion rate.

The ODR characterises the soil O_2 condition.

Factors influence the ODR

1. Depth of soil, temperature, moisture content, soil texture etc.

    - ODR decreases with the depth of the soil indicating that with an increase in depth the concentration of O_2 decrease.

    - Temperature affects ODR by increasing rate of respiration and diffusion co-efficient of O_2 in water, the solubility of O_2 in liquid phase is decreased.

    - The ODR decreases as the moisture content increases because of filling of pore spaces with moisture resulting no space for gaseous exchange.

The ODR has some practical utility like relationship between plant roots and soil, soil moisture and aeration status and biological activities in the soil etc. which ultimately influence the plant growth by affecting various physical and chemical properties of soil. It has been found that the root growth ceased when the ODR dropped to about 20 g x $10^{-8}/cm^2/min$.

Fick's law : Diffusion is a function of the concentration gradient, the diffusion coefficient of the medium, and the cross sectional area

$$dQ = DA \left(\frac{dc}{dx}\right) dt$$

where dQ = Mass flow (moles) during the time at cross area A (sq cm) $\frac{dc}{dx}$ the concentration gradient (moles/cc(cm)

D = proportionality constant or diffusion coefficient (sq cm/sec)

D - Depends on the property of the medium as well as the gas. It varies directly with square of the absolute temperature and inversely with total pressure. The diffusion coefficient of O_2

is about 1.25 times that of CO_2. The rate of diffusion of CO_2 and O_2 in air is nearly 10,000 times greater than in water.

Characterisation of soil aeration

Various parameters can be used for characterizing soil aeration

1. The volume percentage of soil air (pore space filled with air). This is generally determined by applying the tension equivalent to a water column of 50 cm to a saturated soil on a tension table.

2. Gaseons composition

3. The oxygen diffusion rate (ODR) determined by using platinum micro electrode technique where the diffusing oxygen is allowed to reduce at the platinum electrode at a given electric potential. The rate of diffusion of oxygen to the platinum electrode is used as an index of the rate of diffusion of oxygen through the water film to the roots.

4. Oxygen reduction potential, indicating the oxidized or reduced condition of the soil.

5. Composition of the soil for its reduced components.

Oxidation – Reduction (Redox) Potential

Important chemical characteristic of soils related to soil aeration is the reduction and oxidation states of the chemical elements in these soils. If a soil is well aerated, oxidized states such as those of ferric iron (Fe^{3+}) manganic manganese (Mn^{4+}) nitrate (NO^-_3) and sulphate (SO_4^{2-}) dominate. In poorly drained and poorly aerated soils the reduced form of such elements are found, for example ferrous iron (Fe^{2+}), manganous manganese (Mn^{2+}), ammonium (NH_4^+) and sulphides (S^{2-}). The presence of these reduced forms is an indication of restricted drainage and poor aeration.

An indication of the oxidation and reduction states of systems (given by the oxidation – reduction potential or redox – potential

(Eh). It provides a measure of the tendency of a system to reduce or oxidize chemicals and is usually measured in volts or millivolts. If Eh is positive and high, strong oxidizing conditions exist. If it is low and even negative, elements are found in reduced forms. Oxidised and reduced forms of certain elements in soils and the Redox Potential (Eh) at which change in forms commonly occurs.

Oxidised form	reduced form	Eh at which change of form occurs volts
O_2	H_2O	0.38 to 0.32
NO_3^-	N_2	0.28 to 0.22
Mn^{4+}	Mn^{2+}	0.28 to 0.22
Fe^{3+}	Fe^{2+}	0.18 to 0.15
SO_4^{2-}	S^{2-}	− 0.12 to −0.18
CO_2	CH_4	− 0.2 to −0.28

Importance of Soil Air in plant growth and biological activities in soil

I. Growth of plants and development of roots : The growth of plants are adversely affected by poor aeration namely (1) the development of plant roots are restricted or inhibited, (2) absorption of water and nutrients is decreased, (3) the formation of toxic substances are encouraged by poor aeration, especially under puddle waterlogged condition in case of rice cultivation.

II. Adaptation of plant: The ability of different plant species to grow in soil with low soil air porosity varies greatly. Certain plants such as a rice that area adopted to growth without an external source of oxygen for the roots have large internal air spaces. Plants adapted to poor aeration show shallow root system in the upper part of the soil where aeration is greatest.

III. Microbial population and activity: The poor soil aeration decreases the microbial activity as well as oxidation of organic matter. The beneficial biological process carried out by different O.M. such as decomposition of O.M. biological nitrogen fixation, nitrification etc. largely depends upon the aeration status of soil.

IV. Production of toxic substance : Heavy rainfall, excessive irrigation or puddling together with flooding brings a reduced condition of the soil and subsequently reduction of elements like Fe and Mn and slow rate or incomplete decomposition of OM will take place in the soil. Thus production of various organic toxic acids like lactic, butyric and citric acids etc. and also toxic concentration of Fe and Mn are found in the anaerobic soil condition, absence O_2. Due to production of such toxic substance plant root get injury and thereby lost their ability to absorb nutrients and water for their growth.

V. Absorption of water and nutrients : Effects of aeration on the uptake of nutrients by plants may result from changes in nutrient availability that occur in the soil in response to the aeration or they may result from changes in metabolic status of the plants. Micronutrients like Cu & Zn show deficiency under poor aeration (water logged) conditions. Poor aeration decreases the uptake of water, as evidenced by the wilted condition of many plants after flooding. The permeability of roots to water decreases under condition of poor aeration.

VI. Incidence of diseases : Plant roots and soil-borne disease organisms occupy the same environment.

SOIL TEMPERATURE AND COLOUR

Sources of soil heat

There are various sources of soil heat namely solar radiation, biochemical reactions, conduction, precipitation, exposure and vegetation etc.

1. Solar radiation

Radiant energy from the sun is the power source that determines the thermal regime of the soil and growth of plants. In presence of solar energy an adequate water supply and sufficient plant nutrients to maintain plant growth. Radiation from the sun is distributed by components of the earth's atmosphere as it passes downward to ward the earth. The heat absorbed by the surface of the earth from the solar radiation is affected by several parameters like latitude, distribution of land and water, slope of the land etc.

The angle at which the sun's rays meet the earth greatly influences the amount of radiation received per unit area. The radiation received per unit area decreases with an increase in the angle.

The angle at which the rays of the sun strike a steep south slope is entirely different from that on a steep north slope. The southern slope received more solar radiation per unit area. The temperature of the soil is always higher on southern exposures than on northern.

The pressure of large amount of water in soil tends to stabilize the temperature because of the high specific heat of water, which is responsible for the absorption of large amount of heat.

Solar radiation is the greater and main source of soil heat.

2. Bio-chemical reactions : In the soil atmosphere a variety of chemical reactions are going on and during such reaction liberation of large amount of heat in the soil environment results. Decomposition of organic matter and other crop residues in the soil and other microbial processes liberate large amounts of heat in the soil and thus contributes soil heat.

3. Conduction: The inner atmosphere of the earth is very hot, the conduction of heat to the soil environment is very slow. Generally during night, the uppermost surface soil becomes cooler than subsurface soil. Thus heat flows from the regions of sub soil to the region of surface soil (cooler soil layer).

4. Precipitation : During the winter season precipitation increases soil heat because of its (precipitation) higher specific heat.

5. Exposure : Exposure is the little importance in the tropics because of the high elevation of the sun. It is of significance in the middle latitude where the elevation is lower. The greater the percentage of diffuse sky radiation in the global radiation, the smaller is the difference in the incoming solar energy per unit area for slopes of different exposures.

6. Vegetation : Vegetation plays a significant role of soil heat because of the insulating of properties of plant cover. Bare soil is unprotected from the direct rays of the sun and becomes very warm during the hottest part of the day.

The major impact of vegetation are associated with (1) the albedo effect (absorptivity of soil), (2) decreasing the depth of penetration of global radiation through the canopy, (3) increasing the heat in evapotranspiration (4) decreasing the rate of heat loss from the soil through its insulating influence.

Loss of Heat : There are various factors that influence the loss of soil heat like radiation, conduction, evaporation and precipitation.

1. Radiation : The quantity of heat is absorbed by the surface soil does not remain constant. Some portion of the soil absorbed soil heat is lost to the atmospheric environment by radiation.

2. Conduction : The conduction process means the transmission of anything from one point to the other point. Here heat is transmitted to the sub-surface horizon of the soil from the surface soil layer by the conductance process and thereby results the loss of soil heat from the surface soil.

3. Evaporation : The major portion of the global radiation in humid climates is used in the process of evapo-transmission. This process consume 580 Cal/g of water that is changed from the liquid to the gaseous phase. This heat energy is lost to the soil and ultimately returned to the atmospheric air resulting loss from the soil mass. This results a cooling effect especially at the surface.

    The potential evapotranspiration process may be defined as the amount of water that will be lost by evaporation and transpiration from a surface that is completely covered with vegetation if there is sufficient water in the soil at all the times for the use of the vegetation.

4. Precipitation : During summer months, precipitation has generally a cooling action in the soil, because rainwater usually have a lower temperature than soil.

Factors Influence of Soil Temperature

1. Composition of the soil – Soils containing much more mineral matter get heated very easily than those of soils containing higher amount of O.M.

2. Soil structure – by controlling pore spaces spheroidal type of structure warm up more quickly.

3. Soil texture – A light-textured sandy soil, absorb heat very quickly than heavy textured clayey soils. A heavy soils carry a great quality of water and due to this reason it warms up very slowly.

4. Soil moisture – Controlling thermal regime. The specific heat of water is than the soil. Moist soils have a higher specific heat than dry soil. Consequently a moist soil has a lower temperature than dry soil. Moist soil gets heated very slowly and it is cooler than dry soils.

5. Soil colour – Dark colour of soil is due to presence of large amount of humus.

6. Vegetation

7. Irrigation and drainage – Irrigation raises the humidity of the air lowers the temperature over the soil and reduces the daily soil temperature variations. Drainage decreases the heat capacity of wet soils, which raises the soil temperature. This plays an important role in warming up the soil in the spring.

8. Topography –The temperature of the ridged field is higher than those that are level.

9. Compactness of the soil

10. Climate

11. Season

Season : Seasonal variation in a year – Temperature vary summer months in the northern hemisphere like midday, represents, the peak of the global radiation and max. temperature. The winter months daily night temperature lower. The temperature of surface soil is always higher than the air temperature.

Specific heat, Heat Capacity, Thermal Conductivity and Thermal diffusivity.

1. Specific Heat : Specific heat may be defined as the amount

of heat required to raise the temperature of one gram of a substance by 1°C. The specific heat of dry soil (0.2 Cal/g) is only about one fifth that of water (1 Cal/g). Hence moist soil are cooler due to their high specific heat and also due to the heat energy spent in evaporation of soil moisture. One can calculate the specific heat of a soil Cs from the summation of the specific heat times the mass of the individual constituents

$$C_s = C_1M_1 + C_2M_2 + C_3M_3 + C_4M_4 + C_nM_n \text{ (Cal/g°C)}$$

Where C_s – Specific heat of soil

M_1, M_2, M_3, M_4 & C_1, C_2 ——C_nM_n specific heat of individual constituents respectively.

2. Heat Capacity : The heat capacity of a given material is equal to its specific heat multiplied by its mass.

Sp. heat of soil vary with various soil constituents. It is found that quartz has the lowest sp. ht. of the major soil constituents and humus has the highest, excepting water. It is also evident that the heat capacity is influenced by humus and water content of the soil. The heat capacity of a soil constituent is equal to its specific heat times, its density. The heat capacity of the soil per unit volume can be computed by the following equations as

$$C_s - X_s\,C_s + X_w\,C_2 + X_a\,C_a \text{ (Cal/cc (°C)}$$

Where C_s is the heat capacity of the soil, $X_s\,X_w$ and X_a are the volume fraction of the solid mass, water and air respectively.

$C_s\,C_w$ and C_a are the heat capacities of their respective constituents.

Since the solid mass of the soil consists of mineral and organic matter whose heat capacities per unit volume are approximately 0.45 and 0.60 respectively and value of air component is too small to be significant. So the above equation can be simplified as follows :

C_s = 0.45 X_m + 0.60 X_o + X_w ($X_a C_a$ is too small and neglected) where X_m and X_o are the volume fractions of mineral and organic matter respectively.

Thermal Conductivity and Diffusivity

Once heat is gained (solar radiation) by the soil, the soil temperature is governed by the thermal characteristics of the soil i.e. heat capacity, thermal conductivity and thermal diffusivity. The amount of heat required to produce a given change in the soil temperature depends upon the mass and the nature of the soil.

Influence of Soil Temperature on Plant Growth

1. Germination of seeds : If the temperature is too low, the seed fails to germinate or germinate at a slow rate. If temperature high may be injured to seeds.

2. Physical properties of soil : Soil structure influenced by soil temperature on aggregation of the soil as well as on the binding materials present in it.

3. Microbial activity : Microorganisms having thermophobic and thermophilic nature variation in temperature. Microbiological process like mineralization of nitrogen, nitrogen fixation, pesticide degradation etc. are influenced by the temperature.

   Temperature below 5°C – Activity of M.O's is lowest

   and above 45 ° C – Activity of M.O's is lowest

   25-35 ° C optimum temperature for the activity of M.O.

4. Decomposition of O.M. in the soil

   Low soil temperature – decomposition low – toxic materials developed high temperature – fast decomposition – beneficial products of O.M.

5. Absorption of water

6. Availability of nutrients

7. Root growth soil temperature low temperature encourages white succulent roots with little branching, while high temperature encourages a browner, finer and much more freely branching root system.

8. Plant diseases – low soil temperature – weakly parasitic fungi will grow actively kill the seedlings.

SOIL COLOUR

Soil colour indicates many soil features. A change in soil colour from the adjacent soils indicates a difference in the soil's mineral origin (parent material) or in the soil development. Soil colour varies among different kinds as well as within the soil profile of the same kind of soil. It is important soil properties through which its description and classification can be made.

Kinds of soil colour

Soil colour is inherited from its parent material and that is referred to as lithochromic e.g. red soils developed from sand stone. Besides soil colour also develops during soil formation through different soil forming processes and that is referred to as acquired or pedochromic colour red soils developed from granite, gneiss or schist.

Factors affecting soil colour

There are various factors or soil constituents that influence the soil colour which are as follows -

1. Organic matter : Soil containing high amount of organic matter show the colour variation from black to dark brown.

2. Iron compounds soils containing higher amount of iron compounds generally impart red, brown and yellow tinge colour.

3. Silica, lime and other salts – Some times soils contain either large amounts of silica and lime or both – due to presence of such materials in the soil the colour of the soil appears like white or light coloured.

4. Mixture of organic matter and iron oxides very often soil contain a certain amount of organic matter and iron oxides. As a result of their existence in soil the most common soil colour is found and known as brown.

5. Alternate wetting and drying condition: During monsoon period due to heavy rain the reduction of soil occurs and during dry period the oxidation of soil also takes place. Due to development of such alternating oxidation and reduction condition, the colour of soil in different horizons of the soil profile is variegated or mottled. This mottled colour is due to residual products of this process especially iron and manganese compounds.

6. Oxidation reduction conditions : When soils are water logged for a longer period, the permanent reduced condition will develop. The presence of ferrous compounds resulting from the reducing condition in waterlogged soils impart bluish and greenish colour.

    Therefore, it may be concluded that the soil colour indirectly indicative of many other important soil properties. Besides soil colour directly modify the soil temperature e.g. dark-coloured soils absorb more heat than light-coloured soils.

Determination of soil colour

The soil colours are best determined by the comparison with the Munsell colour. This colour chart is commonly used for this purpose. The colour of the soil is a result of the light reflected from the soil. Soil colour rotation is divided into three parts :

1) Hue   (2) Value   (3) Chroma

1) Hue – it denotes the dominant spectral colour (red, yellow, blue and green).

2) Value – it denotes the lightness or darkness (the amount of reflected light).

3) Chroma – it represents the purity of the colour (Strength of the colour).

The Munsell colour notations are systematic numerical and

letter designations of each of these three variables (Hue, Value and Chroma). For example, the numerical notation YR (Yellowish-red) 2.5 YR 5/6 suggests a hue of 2.5 YR, value of 5 and chroma of 6. The equivalent or parallel soil colour name for this Munsell notation is red.

MUNSELL

SOIL COLOR CHART (Fig.)

1) The Hue – notation of a colour indicates its relation to Red, Yellow, Green, Blue and Purple

2) The Value – notation indicates its lightness, or darkness (the amount of reflected light)

3) The Chroma – notation indicates its strength (purity)

Hue – designated by a symbol in the upper right hand corner of the card (Fig.)

Value – increases horizontally they increase the chroma from left to right (Fig.)

Value – notation of each chip is indicated by the vertical scale in the far left column of the chart (Fig.)

Chroma – notation is indicated by the horizontal scale across the bottom of the chart.

Hue – Colour of the rainbow (R for red, YR for Yellow red, Y for Yellow) preceded by number from 0 to 10.

Hue becomes more yellow and less red as the number increases

The middle of the letter range is at (5) 0 points coincides with 10 point of the next redder hue. Thus 5 YR is in the middle of the Yellow-red hue, which extends from 10 R (zero YR) to 10 YR (Zero Y)

Value – consists of number from 0 for absolute black, to 10 for absolute white hue 5 YR, value 5, Chroma 6, is 5 YR 5/6 a yellowish-red.

Using : holding the soil sample directly behind the apertures – closest matching colour chips

Moist soil sample/dry soil sample

5 YR reddish brown 4/4 : ¾ moist

Other uses washbleeding of 2000

Colour chart can also be used for the evaluation of skin, hair, eye colour in anthropology, criminology, pathology and forensic medicin.

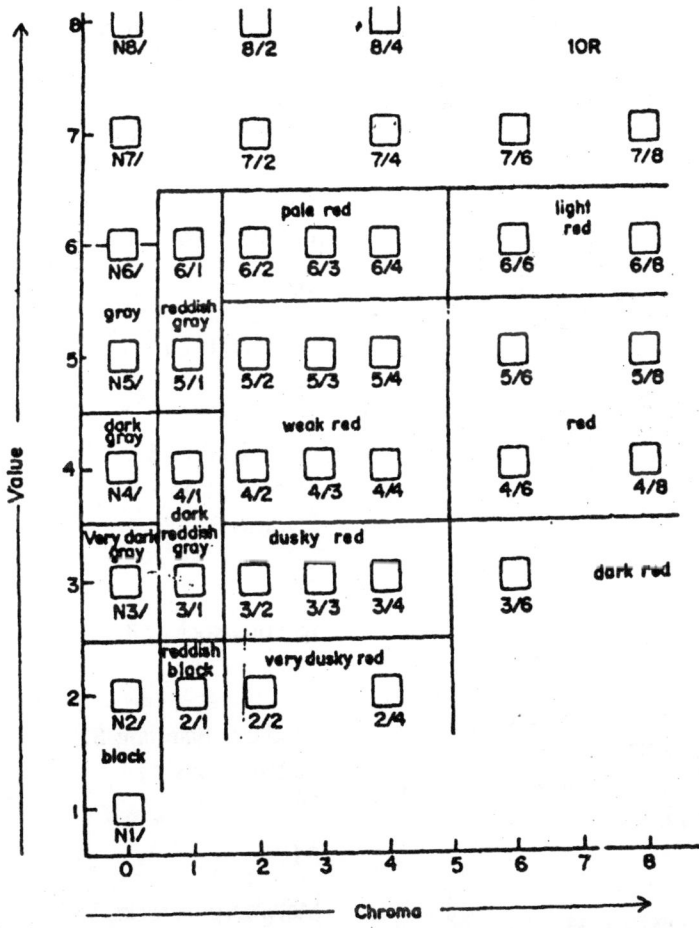

Fig. Page of munsell colour chart shown of several combination of value for hue of 10R

Index